Open Innovation

Open Innovation

Academic and Practical Perspectives on the Journey from Idea to Market

EDITED BY ARTHUR B. MARKMAN

Oxford University Press is a department of the University of Oxford. It furthers
the University's objective of excellence in research, scholarship, and education
by publishing worldwide. Oxford is a registered trade mark of Oxford University
Press in the UK and certain other countries.

Published in the United States of America by Oxford University Press
198 Madison Avenue, New York, NY 10016, United States of America.

© Oxford University Press 2016

All rights reserved. No part of this publication may be reproduced, stored in
a retrieval system, or transmitted, in any form or by any means, without the
prior permission in writing of Oxford University Press, or as expressly permitted
by law, by license, or under terms agreed with the appropriate reproduction
rights organization. Inquiries concerning reproduction outside the scope of the
above should be sent to the Rights Department, Oxford University Press, at the
address above.

You must not circulate this work in any other form
and you must impose this same condition on any acquirer.

Library of Congress Cataloging-in-Publication Data
Names: Markman, Arthur B., editor.
Title: Open innovation : academic and practical perspectives on the journey
from idea to market / edited by Arthur B. Markman.
Description: New York : Oxford University Press, 2016.
Identifiers: LCCN 2015043540 | ISBN 9780199374441
Subjects: LCSH: New products—Management. | Research, Industrial—Management. |
Technological innovations—Management. | Diffusion of innovations.
Classification: LCC HF5415.153 .O64 2016 | DDC 658.4/063—dc23
LC record available at http://lccn.loc.gov/2015043540

CONTENTS

Contributors vii

1. The Process of Open Innovation 1
   Arthur B. Markman

2. Open Innovation Through Strategic Design-by-Analogy 13
   Katherine Fu and Christian Schunn

3. Getting the Most out of Brainstorming Groups 43
   Paul B. Paulus, Jubilee Dickson, Runa Korde, Ravit Cohen-Meitar, and Abraham Carmeli

4. Applying the Creative Problem Solving Process to Open Innovation 71
   Wayne Fisher

5. Opportunity Thinking Approach to Open Innovation: Seeing Clearly the Elephant in the Room 91
   Pam Henderson, Francesca Lorenzini, and Gregory P. Pogue

6. Better, Faster, Safer, and Cheaper: USAA Roof Inspections with Pole Cam 121
   Cliff Zintgraff, Matt Reedy, Shane Osborne, and Rob Pacheco

7. An Open Invitation to Open Innovation: Guidelines for the Leadership of Open Innovation Processes 141
   Diana Rus, Barbara Wisse, and Eric F. Rietzschel

8. Innovation Cells: Company Beachheads in Technology Universities 169
   Felix Cardenas, Tony Davila, and Daniel Oyon

9. Building an Innovation Coral Reef: The Austin Technology Incubator Case Study 203
   Gregory P. Pogue, Keela Thomson, Rosemary French, Francesca Lorenzini, and Arthur B. Markman

Index 225

CONTRIBUTORS

Felix Cardenas
EGADE Business School—
Tecnologico de Monterrey
Monterrey, Mexico
The University of Texas at Austin
Austin, TX

Abraham Carmeli
School of Management
Tel Aviv University
Tel Aviv-Yafo, Israel

Ravit Cohen-Meitar
Tmurot
Tel Aviv, Israel

Tony Davila
IESE Business School
University of Navarra
Barcelona, Spain

Jubilee Dickson
Department of Psychology
University of Texas at Arlington
Arlington, TX

Wayne Fisher
Rockdale Innovation
Cincinnati, OH

Rosemary French
IC^2 Institute
The University of Texas at Austin
Austin, TX

Katherine Fu
Georgia Institute of Technology
Atlanta, GA

Pam Henderson
NewEdge
Richland, WA

Runa Korde
Department of Psychology
University of Texas at Arlington
Arlington, TX

Francesca Lorenzini
IC^2 Institute
The University of Texas at Austin
Austin, TX

Arthur B. Markman
Department of Psychology
The University of Texas at Austin
Austin, TX

Shane Osborne
USAA
San Antonio, TX

Daniel Oyon
Department of Accounting
and Control
University of Lausanne
Lausanne, Switzerland

Rob Pacheco
USAA
San Antonio, TX

Paul B. Paulus
Department of Psychology
University of Texas at Arlington
Arlington, TX

Gregory P. Pogue
IC^2 Institute
The University of Texas at Austin
Austin, TX

Matt Reedy
USAA
San Antonio, TX

Eric F. Rietzschel
Department of Psychology
University of Groningen
Groningen, The Netherlands

Diana Rus
Creative Peas
Amsterdam, The Netherlands

Christian Schunn
Department of Psychology
University of Pittsburgh
Pittsburgh, PA

Keela Thomson
Department of Psychology
UCLA
Los Angeles, CA

Barbara Wisse
Department of Psychology
University of Groningen
Groningen, The Netherlands

Cliff Zintgraff
IC^2 Institute
The University of Texas at Austin
Austin, TX

Open Innovation

1

The Process of Open Innovation

ARTHUR B. MARKMAN ■

Innovation is such a powerful buzzword in business that it is difficult to find any description of a company or speech by a corporate leader that does not tout the company's ability to innovate. The importance of innovation has led to much writing (in the form of books, blogs, and magazine articles) on what makes companies successful as innovators. Much of this work comes in the form of stories of successful new products and services, innovators, and companies. The aim of this work is to provide a model for other organizations that want to improve their ability to innovate.

In practice, however, the specific path that any particular company or region takes to innovation is unlikely to work successfully when transferred directly to another setting. Instead, innovation techniques need to be adapted to the context in which they are going to be applied. Transferring an innovation solution from one case to another requires finding a level of abstraction that describes a set of principles that can be reused.

This book focuses on *open innovation*. Open innovation is a term coined by Chesbrough (2003) to refer to a process that moves innovation beyond the boundaries of a particular organization. Innovation itself is the method that supports the commercialization of new products and business processes. Historically, innovation was done within firms or by inventors who created a new technology and then formed a business around it. This mode of innovation is still common.

Open innovation recognizes that firms (or specific divisions within firms) may not have all of the resources they need to innovate successfully in house. Instead, mechanisms are needed to expand the capabilities of a firm as needed in order to bring new technology and processes to market. In large organizations, this may require allowing input from different divisions within the firm. For all firms, open innovation may also reach outside the firm to make use of expertise from others.

In this chapter, I explore some of the forces in the business environment that favor open innovation. Then, I distill the lessons that these forces have for large organizations. Next, I examine the role of the idea-generation process in innovation. Finally, I discuss ways to put all of the pieces of innovation together into a coherent framework for open innovation. Along the way, I point forward to chapters in this volume that expand on the themes discussed here.

THE FORCES DRIVING OPEN INNOVATION

The rise of Silicon Valley in the 1970s has been a touchstone in discussions of innovation because many of the most successful high-tech firms in the latter half of the 20th century emerged from this region. Saxenian (1996) provides an in-depth discussion of the history of Silicon Valley and the reasons why it surpassed other areas of the United States in the development of new technology companies. As she discusses, in the 1960s, the Boston/Route 128 corridor and the region around Stanford University (which became the Silicon Valley) were two significant hubs of high-tech activity. By the 1980s, Silicon Valley had emerged as the center of high-tech innovation, while the East Coast faded.

In the 1960s, both regions had the ingredients for success in high tech. There was an infrastructure for developing new ideas and bringing them to market. There were important universities that were developing new ideas and providing talented students who could work in the tech sector.

There were sources of financing to develop and market new ideas. Indeed, the high levels of government research funding that went to universities on the East Coast such as the Massachusetts Institute of Technology might have given the Boston area an early advantage in the development of high-tech firms.

However, Saxenian (1996) points out that the two regions differed significantly in the structure of their ecosystems. The East Coast was dominated by large companies such as IBM, Digital Equipment Corporation, and Raytheon. The West Coast had some large companies, such as Hewlett-Packard, but most companies emerged in the form of small start-ups organized around new technologies. Of course, some of these, such as Apple and Cisco, eventually became large corporations.

This difference in ecosystem influenced innovation strategies in these regions. On the East Coast, companies were secretive about their innovations. New technologies were developed within a firm. The aim of projects was for a particular firm to handle all stages of a project from research and development (R&D) to manufacturing and sales.

An interesting side effect of this mode of innovation was that East Coast firms were often conservative about the projects they took on. The career path in large corporations allowed individuals to enter the management track of the companies and to advance through the ranks. However, promotions typically went to those managers whose projects were successful. Because many disruptive innovations fail, there was a disincentive for managers in large companies to support these projects. Instead, incremental innovations with a high likelihood of moderate success were preferred to disruptive innovations with a small chance for significant success.

The West Coast adopted a different model. Most innovations were developed by small firms. Companies were created around new technologies, and teams were assembled to bring these developments to market. A company might focus on a particular product such as a chip design, network component, or storage device. Because the new entrepreneurial ventures were small, there was pressure on these firms to engage with the rest of the high-tech community. Potential competitors would collaborate on projects when the outcome would benefit both companies. Rival firms were not seen solely in competitive terms.

Because small firms on the West Coast were organized around emerging technologies, there was a willingness to support disruptive innovations. Although many startups failed, this failure was viewed as a part of the entrepreneurial process. Consequently, failure did not have the same negative connotation within Silicon Valley that it did in larger East Coast firms. Individuals who had been part of failed ventures in the past were still included in new projects and were given opportunities to pass on the lessons learned from previous mistakes. This fluid movement of business talent and shifting set of alliances among companies made the ecosystem of Silicon Valley much more open than that on the East Coast.

A number of factors influenced the success of the open model of innovation favored on the West Coast. An important factor is the fast-moving pace of change in the high-tech sector. Moore's law (an observation by the technologist Gordon Moore (1965)) pointed out that there has been an exponential improvement in computing technology. The number of transistors that could be placed on a particular chip doubled every 2 years. More generally, computing technology rapidly gets smaller, cheaper, faster, and more efficient.

Because of this rapid improvement, complex computing devices (that incorporate memory chips, central processing units, storage devices, and networking) require innovations of all of their constituent components as well as innovation in the development of architectures and software that bring these components together in new ways. An open system of innovation allows independent companies to work on each of these components and to optimize them while separate firms can focus on more global hardware projects or key pieces of software. The more unified model on the East Coast struggled because it is difficult for a single firm to be a leader in each of the components of a device.

In addition, small firms on the West Coast could compete with each other to develop new components that were used in larger devices. Consequently, firms developing large devices that used these components had the opportunity to select from many potential vendors whose products offered new features that would enhance the capabilities of the devices they construct.

Because these components were being developed by other companies, the manufacturers of sophisticated systems did not need to take on the risk and cost of R&D associated with optimizing specific components.

Silicon Valley was also able to use the ecosystem to spread the risk associated with innovation across many entities. Small firms whose products did not succeed simply went out of business. The employees of these failed enterprises were often brought into other ventures. Manufacturers whose products were incorporated into larger devices that failed would simply sell their wares to other startups. In addition, capital was raised from a number of sources in the region. Thus, unlike large companies that could be adversely affected by product failures, the collection of startups in Silicon Valley were set up to minimize the catastrophic risk of failure and to spread the wealth in the event of success.

Throughout the years, there have been many regions that hoped to become the next Silicon Valley. In the 1980s, for example, Austin, Texas, hoped it could duplicate the success of Silicon Valley after the MCC Consortium opened a high-tech research facility in the city (Gibson & Rogers, 1994). Although Austin did succeed in becoming a high-tech hub, it did not do so by using the same model as Silicon Valley. The success of Austin involved a combination of many large high-tech companies that moved to the city and brought significant business and technical talent with them combined with a robust startup environment that drew from these companies and the University of Texas. Indeed, a key lesson for open innovation in this book is that regions should not expect to copy an existing model to become a center for innovation. Instead, it is important to understand the local ecosystem to enhance the strengths of that region.

The technopolis model was developed as an attempt to explain the way cities such as Austin grew into centers of open innovation (Smilor, Gibson, & Kozmetsky, 1989). This model suggests that the community (including government agencies) needs to support the activities of both large corporations and entrepreneurial ventures. In addition, large research universities need to provide technical and business expertise to the community. The interactions among these entities create the conditions for the flow of ideas that allows new businesses to thrive. The triple helix model provides a different perspective on this issue, but it also suggests that the government, university, and industry are crucial players in the development of a thriving entrepreneurial culture (Etzkowitz & Leydesdorff, 1998; Leydesdorff & Etzkowitz, 1996).

A number of structures have been created that try to take advantage of the local community. For example, networked incubators are organizations that facilitate the growth of new ventures by embedding them in the social structure of a region (Bollingtoft & Ulhoi, 2005; Hansen, Chesbrough, Nohria, & Sull, 2000).

To explore structures such as this in more detail, Chapter 8 explores a region that created a number of successful open innovation projects. The stories of these centers are distilled into lessons for how to organize a community to have a positive influence on innovation.

LESSONS FOR LARGE COMPANIES

Interestingly, at the same time that Silicon Valley was churning out successful high-tech startups, advances in supply-chain management supported unprecedented growth of large manufacturers and retail outlets. Entrepreneurs who manufactured and sold consumer products were being displaced by large companies that could produce products faster and more cheaply and deliver those products to consumers at lower prices. Initially, this revolution focused on outlets for consumer products, but by the 1990s, manufacturers such as Dell also turned computers into inexpensive commodities.

Historically, of course, large companies have been great sources of R&D. Bell Labs was born out of the telecommunications monopoly of the Bell System. It was an important source of new ideas in the telecommunications area. Because Bell was a monopoly, however, it could control the pace of the introduction of new technologies.

Drug companies have significant infrastructure for developing new compounds and for conducting the arduous testing required to get new drugs through the approval process. Thus, the large drug companies continue to be important players in the creation of new pharmaceuticals.

However, the industries in which corporate R&D succeeds are those in which discoveries occur through a slow process of discovery.

One lesson that could be drawn from the success of Silicon Valley is that large companies are not suited to the rapid pace of innovation in the high-tech era. That is, despite the success of large firms at bringing existing products to market cheaply, and their ability to do large-scale research, perhaps only small companies are nimble enough to be a sustainable source of innovation in the modern era.

One purpose of open innovation is to bring more of the feel of the small company ecosystem into larger organizations. A classic case of open innovation in a large company is Connect & Develop, which was created by Procter & Gamble (P&G; Dodgson, Gann, & Salter, 2006). In Connect & Develop, P&G publishes the specifications for a component that it needs for a product and it offers a payment that will be given to the first team that develops that component. For example, the desired component might be a chemical compound with particular properties. Individuals and companies can access the Connect & Develop website searching for opportunities. The first entity that meets the specifications laid out by P&G gets paid. In this way, P&G benefits from a large pool of development teams, but it only has to pay the team that is successful.

Connect & Develop is just one method for opening up the innovation process. There are many different techniques for open innovation. Many of these can help large companies benefit from the creation of ecosystems that make the large organization more nimble. These approaches are also valuable for regions that are seeking to improve the innovation capabilities of companies located there. Despite the importance of open innovation, however, there are many pitfalls that organizations must avoid. These pitfalls are explored in Chapter 7.

INNOVATION AND IDEATION

As important as the community is for the success of a new venture, disruptive technologies and processes start with great ideas. One of the benefits of opening up the process of innovation is that it allows a wider range of individuals to develop new ideas.

Traditionally, organizations had R&D teams that were given the task of creating and refining new ideas. In addition, specific business units might take the time to brainstorm periodically to move that unit forward. This approach to idea generation limits the number of people who participate in the process.

One thing we know about idea generation, however, is that groups that generate many ideas are the ones most likely to generate good ideas (Linsey et al., 2011; Paulus & Nijstad, 2003). Thus, anything that limits the sources of ideas in a group will also cut down on the effectiveness of idea generation.

There are several ways that idea generation might be limited. One is through the techniques that are used to create ideas. The term brainstorming, for example, refers to a specific technique for idea generation developed by Osborn (1957). This technique involves having people in a group state as many ideas as they can while other group members build on those ideas. Unfortunately, decades of research demonstrate that this widely used technique actually limits the number of ideas (and the number of good ideas) that are generated by a group (Mullen, Johnson, & Salas, 1991).

There are many ways to expand the process and get more people involved in idea generation without suffering from the limitations of brainstorming. One way large companies can succeed at opening up the idea generation process is to create innovation facilities with trained innovation leaders to help groups maximize their effectiveness. Chapter 4 describes an innovation center at P&G called the GYM and explores some of the idea generation techniques that were used there. Large companies such as P&G have tried to adopt methods of design thinking firms such as IDEO and to bring those facilities in house to improve innovation. Although these corporate centers have helped improve the way large companies innovate, many of them (including the GYM) have ultimately been shut down because it is difficult to measure the financial impact of these centers on a large company's bottom line.

Group idea generation can also be improved by helping group members to build on each other's ideas. This social element allows more people to ensure that the strengths and weaknesses of an idea are explored as fully as possible and that the resulting ideas reflect a broad consensus on the best way to proceed with a project. Chapter 3 examines this process in-depth and examines methods that allow brainstorming groups to maximize the contribution of all group members.

Another way to open up the innovation process is to bring in expertise from domains that do not seem obviously relevant from the way the problem is stated at first. Often, people try to solve problems within the domain in which they were stated. Designing new integrated circuits, for example, would focus primarily on expertise in the domain of semiconductors. However, many innovations emerge from analogies drawn between the problem and other domains of knowledge (Basalla, 1988; Christensen & Schunn, 2007; Markman, 2012; Polya, 1945). Techniques that improve the use of analogies in idea generation open up innovation by expanding the range of information that is considered relevant to the problem being solved. The role of analogy in idea generation is explored in Chapter 2.

These techniques for improving idea generation reflect a growing appreciation that an understanding of psychology can benefit corporate effectiveness. The importance of the behavioral and social sciences is at the core of the design thinking approach taken by IDEO (Kelley & Littman, 2001). Authors such as Malcolm Gladwell, Dan Pink, and Danny Kahneman have brought research in the behavioral sciences to a mainstream business audience in ways that have reinforced that many aspects of psychology are simply not obvious when one introspects about one's own behavior.

Idea generation is one of those places where the things that seem intuitively obvious (e.g., brainstorming) can go wrong. Thus, Chapters 2, 3, and 4 that explore science-based techniques for ideation are also a good case study for how psychological research can benefit business practice.

PUTTING THE PIECES TOGETHER

Within an organization, it is crucial to evaluate ideas to ensure that they are worth pursuing. There are many examples of ideas and technologies that were commercialized without any sense that there was a market for them. For example, the Segway is a technological marvel, using sophisticated technology to keep the device upright allowing individuals to travel short distances quickly. As fascinating as the device is as a technology, the hype surrounding the release of the Segway was even more impressive. Yet, despite the news coverage surrounding the product launch, the Segway was largely a flop. It has been purchased primarily by tour companies and by those who need to walk for long periods of time, such as shopping mall security guards.

One problem with the evaluation process in many organizations is that it does not look broadly enough to determine whether the idea has a real chance to succeed. Techniques of open innovation are also useful for broadening the range of people and groups who take part in the evaluation of an idea. One core way to set the criteria for the evaluation of a new idea is to focus on the opportunities that the process, technology, or product will address. The role of these opportunities in open innovation is explored in Chapter 5.

Finally, academic discussions of innovation can often dig so deeply into the specifics of cases that it is not clear how they generalize, or they can focus on elements of innovation that are so abstract that it is not clear how to apply them in real situations. For this reason, several of the chapters in this book have been written by (or in collaboration with) practitioners of open innovation. In addition to some of the chapters previously mentioned, Chapter 6 presents a comprehensive case study of the development of an idea at USAA, a large insurance company. This chapter follows the idea from initial generation to evaluation and through a process of redesign when the initial idea did not work as desired. Key to this example is the number of different units within the company that were given an opportunity to influence the idea as it matured.

USING OPEN INNOVATION

There are two aims of this book. The first is to provide a source for researchers who want to continue to explore the elements of open innovation. These chapters provide a snapshot of research across a range of aspects of the innovation process. These chapters reflect contributions from researchers across many different disciplines. We hope to inspire people from research areas that have not focused on innovation in the past to devote their research to the study of this crucial topic.

In addition, this book has important lessons for people engaged in the practice of innovation in communities and organizations. These chapters provide a number of important principles that can be used to help companies to improve their innovation capacities and to allow regions throughout the world to duplicate the success of areas such as Silicon Valley and Austin by engaging similar principles rather than by copying specific aspects of these regions directly.

References

Basalla, G. (1988). *The evolution of technology.* Cambridge, England: Cambridge University Press.

Bollingtoft, A., & Ulhoi, J. P. (2005). The networked business incubator—Leveraging entrepreneurial agency. *Journal of Business Venturing, 20*(2), 265–290.

Chesbrough, H. W. (2003). *Open innovation: The new imperative for creating and profiting from technology.* Cambridge, MA: Harvard Business Review Press.

Christensen, B. T., & Schunn, C. D. (2007). The relationship of analogical distance to analogical function and pre-inventive structure: The case of engineering design. *Memory and Cognition, 35*(1), 29–38.

Dodgson, M., Gann, D., & Salter, A. (2006). The role of technology in the shift towards open innovation: The case of Procter & Gamble. *R&D Management, 36*(3), 333–346.

Etzkowitz, H., & Leydesdorff, L. (1998). The endless transition: A "triple helix of university–industry–government relations. *Minerva, 36*(3), 203–208.

Gibson, D. V., & Rogers, E. M. (1994). *R&D collaboration on trial.* Boston, MA: Harvard Business School Press.

Hansen, M. T., Chesbrough, H. W., Nohria, N., & Sull, D. N. (2000). Networked incubators: Hothouses of the new economy. *Harvard Business Review, 78*(5), 74–84.

Kelley, T., & Littman, J. (2001). *The art of innovation: Lessons in creativity from IDEO, America's leading design firm.* New York, NY: Crown Business.

Leydesdorff, L., & Etzkowitz, H. (1996). Emergence of a triple helix of university–industry–government relations. *Science and Public Policy, 23*(5), 279–286.

Linsey, J. S., Clauss, E. F., Kurtoglu, T., Murphy, J. T., Wood, K. L., & Markman, A. B. (2011). An experimental study of group idea generation techniques: Understanding the roles of idea representation and viewing methods. *Journal of Mechanical Design, 133*(3). doi:10.1115/1.4003498

Markman, A. (2012). *Smart thinking.* New York, NY: Perigee Books.

Moore, G. E. (1965). Cramming more components onto integrated circuits. *Proceedings of the IEEE*, *86*(1), 82–85.

Mullen, B., Johnson, C., & Salas, E. (1991). Productivity loss in brainstorming groups: A meta-analytic integration. *Basic and Applied Social Psychology*, *12*(1), 3–23.

Osborn, A. (1957). *Applied imagination*. New York, NY: Scribner.

Paulus, P. B., & Nijstad, B. A. (Eds.). (2003). *Group creativity: Innovation through collaboration*. New York, NY: Oxford University Press.

Polya, G. (1945). *How to solve it*. Princeton, NJ: Princeton University Press.

Saxenian, A. (1996). *Regional advantage*. Cambridge, MA: Harvard University Press.

Smilor, R. W., Gibson, D. V., & Kozmetsky, G. (1989). Creating the technopolis: High-technology development in Austin, Texas. *Journal of Business Venturing*, *4*, 49–67.

2

Open Innovation Through Strategic Design-by-Analogy

KATHERINE FU AND CHRISTIAN SCHUNN ■

A promising basis for the next engines of design innovation practice in open innovation is design-by-analogy, in which designers are exposed to and map design solutions from other domains to the design problem at hand. Although this methodology could be incredibly fruitful for guiding engineering product design processes, there is currently no practical, efficient, procedural way to find good analogies from the sea of potentially millions (or more) of possible sources. This chapter explores the effects of different types of analogical stimuli on design output quality of individual engineering designers to build a better understanding of design-by-analogy from a cognitive perspective. Then a new computational algorithm is developed and tested for exploration of inherent structural forms within a repository of existing design solutions. This new approach provides insights regarding the interrelatedness of different solutions that are both novel and meaningful to engineering designers.

BACKGROUND AND RELEVANT PRIOR RESEARCH

Design-by-Analogy

The use of analogy during design has been frequently studied by researchers in both cognitive science and engineering design. This research has examined how the introduction of analogies affects the ideation process and outcomes (Christensen & Schunn, 2005; Dahl & Moreau, 2002; Goldschmidt & Smolkov, 2006; Linsey, Wood, & Markman, 2008), with some studies unpacking how analogies with different levels of applicability or distance to the current design problem influence designers (Christensen & Schunn, 2007; Tseng, Moss, Cagan, & Kotovsky, 2008).

Interestingly, the benefits of analogy depended on whether the designers had "open goals" (i.e., unsolved problems) in mind when exposed to information that could be relevant to the design problem (Moss, Cagan, & Kotovsky, 2007; Moss, Kotovsky, & Cagan, 2007). For example, Tseng et al. (2008) found that giving individuals information that was analogous but distantly related to the design problem caused them to produce more solutions with a wider range of solution types and higher level of novelty when open goals existed; in the absence of open goals (i.e., prior to the introduction of the problem to be solved), more highly similar analogous information was more beneficial.

Researchers have also explored negative factors associated with introducing analogical information or examples and particularly design fixation (Chrysikou & Weisberg, 2005; Jansson & Smith, 1991; Purcell & Gero, 1996; Smith & Blankenship, 1991), or the "blind adherence to a set of ideas or concepts limiting the output of conceptual design" (Jansson & Smith, 1991). Jansson and Smith showed that introducing examples can cause designers to generate solutions that mimic the examples, including features that specifically violate the design problem objectives. Furthermore, Ward et al. found that designers often included aspects of examples in their solutions, even when explicitly told not to do so, suggesting that designers have little control over the extent to which they are influenced by provided examples (Marsh, Ward, & Landau, 1999; Smith,

Ward, & Schumacher, 1993). In the work presented here, the expectation is to uncover inspirational analogical information that could be useful to the designer if introduced at an appropriate point in ideation (e.g., at the time of open goals) and in a helpful format to increase inspiration but avoid fixation.

ANALOGICAL DISTANCE

A large area of exploration in the study of design-by-analogy is analogical distance. Often, analogical distance is conceptualized as dichotomous, with analogies being either near field or far field. "Near" generally refers to analogies from the same or similar domain as the design problem, whereas "far" generally refers to analogies from a different domain. Likely more relevant to their differential benefits, near-field analogies share more surface features with the design problem than do far-field analogies. Far-field analogies, however, can have functional similarities to the design problem that make them nonetheless apt for generating effective solutions and, by virtue of distance, perhaps previously unexplored solutions in the current problem domain. Some researchers argue that far-field analogies are the most promising for creative insights (Gentner & Markman, 1997). Dahl and Moreau (2002) found that the number of far-field analogies used by designers during ideation is positively correlated with the originality of proposed solutions, as rated by a sample of potential customers. Wilson, Rosen, Nelson, and Yen (2010) found that exposure to surface dissimilar design examples increased idea novelty relative to using no examples, and exposure to surface similar examples decreased variety of ideas relative to surface dissimilar examples. Many design consultants even recommend the use of random analogical stimuli, which can perhaps be thought of as "extremely far-field," to make free or wild associations when ideating (Dyer, Gregersen, & Christensen, 2011). Yet other research argues against the benefits of far-field analogies (Dunbar, 1997; Weisberg, 2009). At a process level, the relevance of far-field analogies to a design problem is often more difficult to comprehend (Casakin & Goldschmidt, 1999), and they can be difficult to retrieve from memory (Forbus, Gentner, & Law, 1994; Gick & Holyoak, 1980).

Currently, practitioners lack a cohesive theoretical account of the effect of analogical distance on design outcomes to guide practice. One possible reason for this is that there is a lack of consistency in the conceptualization and measurement of analogical distance, making research results difficult to compare. Inherently, analogical distance is a continuum rather than just a simple dichotomy with the near/far categories applied at arbitrary points along the continuum. Therefore, it may be that different prior studies in the literature were examining different points on the continuum from near to far. It is possible that the benefits of analogical stimuli for ideation could vary in a curvilinear manner from near to far, with very near and very far stimuli being less beneficial or even damaging (e.g., due to fixation from low novelty at the near end and time wasting due to low relevance at the far end).

Modality of Analogical Stimuli

In developing a useful theory for the use of analogical stimuli in design, a potential variable of interest is the contrast between pictorial and text-based representations of examples. One possible reason to investigate this contrast of particular relevance to our current focus on analogical distance is that pictorial representations, such as sketches, photographs, and engineering drawings, often contain a higher degree of superficial features compared to text-based representations of the same information. When working with detailed prototypes containing many superficial details, designers tend to retrieve fewer far-field analogies from memory (Christensen & Schunn, 2007). Other researchers, however, argue that pictorial-based representations are better for conceptual design. For example, novice designers presented with sketches of example designs tend to produce more novel and higher quality solution concepts on average relative to being presented with text-based example designs (McKoy, Vargas-Hernandez, Summers, & Shah, 2001). At a pragmatic level for the creation of design-by-analogy tools, a particular representation format for potential analogies must be selected; thus, it is important to investigate if it matters whether they are represented in pictorial or text-based formats (Linsey, Murphy, Markman, Wood, & Kortoglu, 2006; Linsey et al., 2008). In addition, it is possible that the effects of example analogical distance vary across text and pictorial formats.

Commonness of Analogical Stimuli

Another variable of high relevance to conceptualizing analogy distance effects is the commonness of example designs (i.e., how common the designs are found in designers' worlds). Near examples might be more common than far examples, even though the two dimensions can vary orthogonally (i.e., one can select near uncommon, near common, far uncommon, and far common examples from which to analogize). Designers are more likely to already have seen and thought about more common example designs, and thus the relevant benefit of being given common designs might be low. However, more common example designs may more effectively activate relevant prior knowledge in a designer, from prior personal exposure to instances, from engineering coursework, or from professional design experiences (Purcell & Gero, 1992).

Research on creativity and problem solving suggests that common prior experiences with an artifact might negatively influence one's ability to flexibly re-represent—use it and combine it with other concepts in a novel manner. Consider, for example, Duncker's (1945) classic candle problem, in which the task is to fix a lighted candle on a wall in such a way that the candle wax will not drip onto a table below, and the given materials are a candle, a book of matches, and a box of thumbtacks. A good solution involves emptying the box of tacks and using it as a platform for the candle; however, this solution eludes most solvers because it requires recognizing an unconventional use of the box as a platform. When the box is presented to solvers empty (i.e., not being used as a container), with the tacks beside it, solvers are much more likely to find the unconventional solution (Adamson, 1952). Similarly, in Maier's (1931) two-string problem, in which the task is to tie two strings together that are hanging from the ceiling just out of arm's reach from each other using various objects available (a chair, a pair of pliers, etc.), people often fail to recognize the solution of tying the pair of pliers to one string and swinging it like a pendulum and catching it while standing on a chair between the strings. These findings demonstrate the phenomenon of "functional fixedness," in which individuals have difficulty seeing unusual alternative uses for common artifacts.

Another relevant finding regarding the effects of example commonness is that individuals who acquire experience with classes of information and procedures tend to represent them in relatively large, holistic "chunks" in memory, organized by deep functional and relational principles (Chase & Simon, 1973; Chi, Feltovich, & Glaser, 1981; Chi & Koeske, 1983). This ability to "chunk" is an important element of expertise and skill acquisition (Anderson & Schunn, 2000; Chase & Simon, 1973; Newell, 1990). However, if the task at hand requires the individual to perceive or represent information in novel ways—for example, to stimulate creative ideation in design—representation of that commonly experienced information in chunks might become a barrier to success (Kaplan & Simon, 1990; Knoblich, Ohlsson, Haider, & Rhenius, 1999; Ohlsson, 1992).

These findings lead to a hypothesis that less common example designs (whether near or far examples from the current design problem) might present a unique advantage over more common example designs in terms of the potential for stimulating creative ideation. Specifically, it could be that less common examples are more likely to support multiple interpretations and thus facilitate broader search through the space of possible solutions. In addition, given that the commonness of example designs in the world (in practice, curriculum, etc.) is related to its representation in designers' long-term memory (e.g., ease/probability of recall), one could hypothesize that less common examples might confer an advantage in terms of the novelty of solution paths they inspire. The next section presents a study about the effects of example commonness and teases them apart from the effects of example distance.

Computational Design Tools

In addition to the kinds of analogies that are most beneficial and stimulating for designers (near/far, common/uncommon, and text/pictorial), it remains unclear how to find these analogies in an efficient or automatic way. Methodologies have been developed for design-by-analogy, however.

Stone and Wood (2000) created a functional basis in order to provide a universal language to facilitate functional modeling, a useful tool in the ideation process, which has been extended and adapted a great deal, one example of which is a biological functional basis (Cheong, Shu, Stone, & Wood, 2008). This functional basis and language of design work is an important aspect of our work because it is the basis for one of the approaches in the methodology that allows for the exploration of functional interrelatedness of patents and is compared to surface interrelatedness.

The use of patents as design aids is not a new area of research either. The theory of inventive problem solving (TRIZ), a heuristic-based theory that helps designers overcome impasses during ideation, hinges on the idea that a solution to a design problem already exists in the patent literature but perhaps in another field of application (Rantanen & Domb, 2002). TRIZ is composed of 40 inventive principles that are based on the all-inclusive set of mechanisms thought to be behind truly "inventive" patents among hundreds of thousands examined. The principles suggest new ways of conceptualizing or representing a design problem. TRIZ and the functional basis have also been combined to create an axiomatic conceptual design model (Zhang, Cha, & Lu, 2007). Methods have been developed to find the interrelatedness between technologies based on patent citation data and the benefits of tapping into the technology knowledge base created by competitors within a particular design field (Chakrabarti, Dror, & Nopphdol, 1993). Syntactic similarity between patent claims has been explored for the purpose of aiding patent infringement research (Indukuri, Ambekar, & Sureka, 2007). In addition, design repositories other than patent data have also been explored as resources for designers, serving as ways to share and reuse designs to streamline the product design of complex engineering systems (Szykman, Sriram, Bochenek, Racz, & Senfaute, 2000).

A STUDY OF THE EFFECTS OF ANALOGICAL DISTANCE, COMMONNESS, AND MODALITY IN DESIGN

We conducted an initial cognitive study to examine the effects of modality, familiarity, and distance (i.e., near field vs. far field) of an analogy on individual engineering designers (Chan et al., 2011). We measured the effects on a number of different measures of creative output: the number of solutions to a given design problem that were produced, the diversity of those solutions, and a measure of the feature transfer from the analogies given to those solutions. The literature suggested that far-field examples may be more beneficial to ideation performance than near-field examples. Although to our knowledge no prior work has been performed testing the effect of commonness of example on ideation performance, related research literature suggested less common examples may be more beneficial than more common examples. There was no clear indication from previous work regarding the effect of modality of example on ideation performance; thus, this study is more exploratory with regard to that manipulation.

Experimental Design

We conducted a 2 (distance: far field vs. near field) × 2 (commonness: more common vs. less common) × 2 (modality: pictures vs. text) factorial experiment to identify eight different types of analogical patents, only one type of which was presented to a given participant. Participants were given a real-world design problem and were asked to generate solution concepts first briefly without examples, such that they understood the problem. Then, they were given two example patents of a particular type by condition (e.g., only far, common patents in text form) to evaluate the effects of examples on problem solving. To establish whether examples of different types enabled or hindered problem solving, a control group of students executed a similar procedure but received no examples.

Participants

Participants were 153 students (predominantly mechanical engineering undergraduates) from two US research universities, recruited from classes and given either extra credit or $15 compensation. Participants ranged from 20 to 38 years in age. Participants were randomly assigned to one of the nine possible conditions in each class by distributing folders of paper materials prior to students arriving in class.

Design Problem

The problem involved designing a low-cost, easy to manufacture, portable device to collect energy from human motion for use in developing impoverished rural communities (e.g., in India and many African countries). This design problem was selected to be meaningful and challenging to our participants. The problem was meaningful in the sense that real-world engineering firms are seeking solutions to this problem and the problem involves social value; thus, students would be appropriately engaged during the task (Green, Dutson, Wood, Stone, & McAdams, 2002; Green & Wood, 2004; White & Wood, 2010). The problem was challenging in the sense that a dominant or accepted set of solutions to the problem has yet to be developed (so students could not simply retrieve past solutions), but it was not so complex as to be a hopeless task requiring a large design team and very detailed task analysis. This problem was used throughout the work reported in this chapter, both in the cognitive work and in the computational work described later.

Selection of Analogical Stimuli

Eight example patents were selected from the US Patent Database by two Ph.D.-level mechanical engineering faculty based on two sets of criteria: (1) balanced crossing of the analogical distance and commonness factors, such that there would be two patents in each of the four possible combinations, and (2) overall applicability to the design problem, over and above analogical distance and commonness. Each participant in the analogy conditions received two examples of a particular type, roughly balanced across conditions for applicability. The patents for each of the conditions are shown in Table 2.1.

The specific operational guidelines for patent selection were as follows:

> *Distance*: Far-field patents were devices judged to *not* be directly for the purpose of generating electricity, whereas near-field patents were those judged to be directly for the purpose of generating electricity.
>
> *Commonness*: More common patents were devices judged likely to be encountered by members of the target population in their standard engineering curriculum and/or everyday life, whereas less common patents were those judged unlikely to be seen previously by the participants.

With respect to the modality factor, in the picture conditions, participants first received a representative figure from the patent, which typically

Table 2.1. PATENTS FOR EACH CONDITION

	Near-Field	Far-Field
More Common	• Waterwheel-driven generating assembly (*6208037*)	• Escapement mechanism for pendulum clocks (*4139981*)
	• Recovery of geothermal energy (*4030549*)	• Induction loop vehicle detector (*4568937*)
Less Common	• Apparatus for producing electrical energy from ocean waves (*4266143*)	• Accelerometer (*4335611*)
		• Earthquake isolation floor (*4402483*)
	• Freeway power generator (*4247785*)	

SOURCE: Reprinted with permission from *ASME Journal of Mechanical Design*, originally appearing in Chan et al. (2011).

provides a good overview of the device, whereas in the text conditions, participants received the patent abstract.

Experimental Procedure

Participants proceeded through the phases in class using a sequence of envelopes to carefully control timing of the task and exposure to examples across conditions. We wanted to ensure that design examples were received after participants had made some progress in ideation because prior work found that examples were most helpful when received after ideation had already begun (Moss, Kotovsky, et al., 2007; Tseng et al., 2008). The overall time allowed for this task was sufficient to allow for broad exploration of the concept space but not enough to develop particular ideas in-depth, given our focus on the ideation process. A comparison of the procedure across groups is depicted in Figure 2.1. Participants were instructed to generate and record using words and/or sketches as many solution concepts to the design problem as they could, including novel and experimental solutions.

Ideation Metrics

The participants generated 1321 total ideas. To thoroughly explore the range of effects of varying the analogical distance, commonness, and modality of

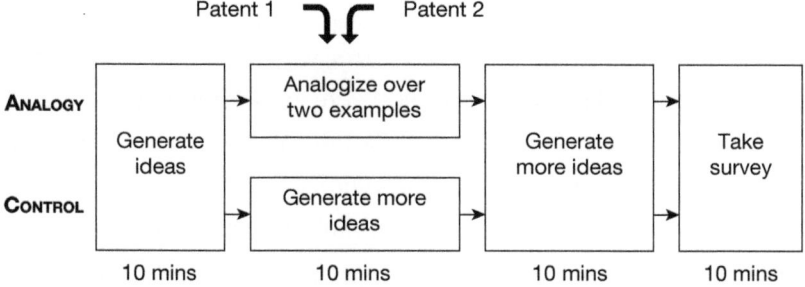

Figure 2.1 Comparison of experimental procedures for analogy versus control groups.
SOURCE: Reprinted with permission from *ASME Journal of Mechanical Design*, originally appearing in Chan et al. (2011).

design examples on conceptual design processes, we applied a range of ideation metrics. The first three metrics—the extent to which solution features were transferred from examples, the quantity of ideation, and the breadth of search through the space of possible solutions—provide measures of the ideation *process* of participants and how they are processing the examples. Examining solution transfer provides insight into the mechanisms by which designers might be stimulated by the examples—for example, whether they actually used solution elements. Measuring quantity of ideation gives us a sense of how participants are exploring the design space—that, is, whether they are generating and refining a small number of ideas or exploring multiple concepts and variations of concepts, which is associated with a higher likelihood of generating high-quality concepts (Terwiesch & Ulrich, 2009). Breadth of search is taken to be a measure of the ability to generate a wide variety of ideas, which is associated with the ability to restructure problems, an important component of creative ability (Boden, 2004; Markman & Wood, 2009; Shah, Vargas-Hernandez, & Smith, 2003).

The final two metrics—quality of solution concepts and novelty of solution concepts—focus on the ideation *products* of participants. We investigate quality because a baseline requirement in design is that concepts must meet customer specifications (Markman & Wood, 2009). We investigate novelty because of the literature consensus that creative products are at least novel (Boden, 2004; Markman & Wood, 2009).

Results and Discussion

Both commonness and distance had effects, which were generally additive effects (Figure 2.2). Far-field examples were associated with higher novelty and more variability in quality of design solutions, which may lead to a greater chance of generating higher quality final design solutions. The analogical transfer that occurred in employing the far-field examples is further than that which is normally required of designers in this type of experimental work—far field often means surface dissimilar, while functionally the same (Blanchette & Dunbar, 2000; Gick & Holyoak, 1980).

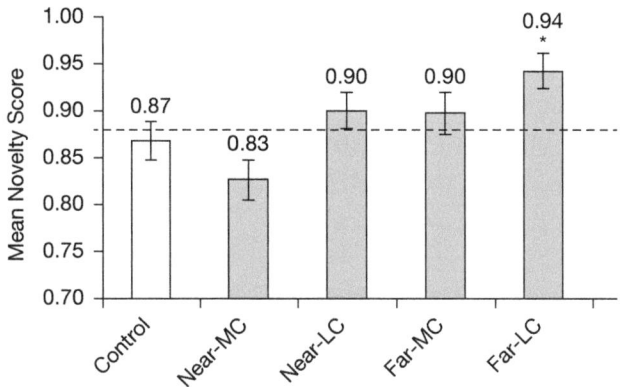

Figure 2.2 Mean novelty of solution concepts by example distance and commonness. $^*p < 0.05$. Error bars are ±1 standard error.
SOURCE: Reprinted with permission from *ASME Journal of Mechanical Design*, originally appearing in Chan et al. (2011).

Here, the far-field examples used were both surface dissimilar and functionally dissimilar in that they were not energy-generating designs, which makes it particularly noteworthy that the participants were able to use and benefit from them. However, the far-field analogies also led to fewer ideas relative to the other two conditions, which may be caused by an increased cognitive load required to process and map the more distant analogical information to the design problem at hand.

Interestingly, there were differences in the positive ideation effects of the far-field and less common examples. Far-field examples led to higher novelty and variability in solution quality but reduced quantity of ideas. Less common examples also led to higher novelty and variability in solution quality, but they additionally led to greater breadth of search of the design space. Thus, far-field examples and less common examples may be beneficial to the designer in different ways: Far-field examples may lead the designer to try ideating narrowly in new areas of the design space (as indicated by less breadth of search and higher solution transfer from the examples to which they were exposed); by contrast, less common examples may benefit the ideation process by causing designers to search many different areas of the design space through new mappings of functions and features from the examples.

The modality of the analogical examples did not change the effects of the distance and commonness manipulations; similar patterns were found across both modalities. However, textual representations resulted in fewer ideas generated relative to exposure to pictorial representations or to no examples. This negative effect may be a result of a potential increased cognitive load involved in processing text compared to processing images.

COMPUTATIONAL SUPPORT FOR DESIGN-BY-ANALOGY

In searching for analogical inspiration from other domains of application, the very large number of design solutions documented in the U.S. patent database serves as a potentially convenient source. However, the size and complexity of the US patent database greatly reduce its usefulness to designers as a source of inspiration. If given a method to automatically extract the interrelatedness and interconnectedness of patents in the space to a given design problem, however, designers might be able to strategically choose which cross-domain designs to expose themselves to or even traverse the space in a more intentional and meaningful exploratory way.

Creating a useful organization of patents is a complex task simply due to the amount of information involved. In addition, it is important to create an organization that corresponds with how humans think about or categorize the information. Therefore, we adapted Bayesian models of human cognition. Bayesian models have been used extensively to describe human cognition, including inductive learning, semantic memory, causal learning and inference, and categorization (Griffiths, Kemp, & Tenenbaum, 2008). Of particular relevance to our challenge, Kemp and Tenenbaum (2008) used Bayesian inference as the foundation for an algorithm that discovers structural form in data. For example, they could discover the color wheel underlying human color similarity judgments, a phylogenetic tree from animal properties, and a simple linear conservative-to-liberal ordering of US Supreme Court justices based on decision patterns. By applying Kemp and Tenenbaum's algorithm to patents, a structural form of the patent space can be discovered, which can uncover insights about how these patents are meaningfully interrelated. Although the focus here is on the patent database, this approach could be applied to any large repository of designs.

Methodology

There are two parts to producing the structures of analogical stimuli presented in this chapter. First, latent semantic analysis (LSA) is used to produce patent similarity data. Second, a Bayesian inference algorithm devised by Kemp and Tenenbaum (2008) is used to discover structural forms in the patent data, using the LSA similarity data as input.

Latent Semantic Analysis

LSA is a computational text analysis tool that builds a semantic space from a corpus of text. This semantic space is then used to compute the similarity between words, sentences, paragraphs, or whole documents for a wide variety of purposes (Deerwester, Dumais, Furnas, & Landauer, 1990; Foltz, Kintsch, & Landauer, 1998; Landauer, Foltz, & Laham, 1998). Note that this semantic space is a high-dimensional vector space (typically 300 or more dimensions, which are the singular values within the singular value decomposition (SVD) step of the implementation) with little inspectable value to humans; the Kemp and Tenenbaum (2008) algorithm is needed to create that inspectable structure. LSA is composed of four main steps:

1. A word-by-document matrix is created from the large corpus of text, in which the columns are the individual text passages (here, the patents), the rows are the words that appear in the documents, and the cells are populated by a tally of the number of times each word appears in each document. For example, the ith column represents patent i, the jth row represents word j, and element ij is the number of times word j occurs in patent i.
2. An "entropy-weighting" step is performed, which is a two-part transformation on the word-by-document matrix that gives a

more accurate weighting of the word-type occurrences based on their inferred importance in the passages. If a word appears very frequently, it is assigned less weight (or importance) through this entropy weighting step (e.g., "a," "the," and "it"). If a word appears very infrequently, it is assigned more importance.
3. SVD is performed on the transformed matrix to create a reduced underlying semantic space that captures regularities of context-based co-occurrence (i.e., two words have a similar representation if they tend to co-occur with other words in similar ways). SVD yields three matrices—U, S, and V, where the rows of U contain the singular vectors corresponding to the words within the LSA space, S is the diagonal matrix of singular values listed in descending magnitude, and the columns of V contain the singular vectors that correspond to each document in the LSA space.
4. The cosine similarity between patent documents can then be calculated by multiplying S and the transpose of V and then calculating the dot product between all pairs of resulting vectors. The result of this final step is a matrix of patent-to-patent similarity values (Deerwester et al., 1990; Foltz et al., 1998; Landauer et al., 1998).

LSA is used in this work to generate similarity data for input into the structural form discovery algorithm.

Structural Form Discovery

The algorithm for discovering structural form as it is applied to the LSA output patent similarity data includes the following steps (Kemp & Tenenbaum, 2008):

1. Preprocess the input similarity data D by shifting the mean of the matrix to zero. Calculate the normalized covariance matrix

for D, defined as $(1/m)DD^T$, where m is the number of features, or nonredundant nontrivial words included in the full set of patents.
2. Find structure S within form F that best captures the relationships between the patents, defined in Bayesian terms as maximizing the posterior probability—the probability that the data have structure S and form F given data D. That is, search for the structure S within form F that jointly maximizes the scoring function $P(S, F|D)$.
3. To identify the structure and form that maximizes the posterior, a separate greedy search is run for each candidate form.
   - Assign all patents to a single cluster.
   - Split a cluster at each algorithm iteration, using a graph grammar for creating structure splits (e.g., adding a branch in a tree).
   - Attempt to improve the score either by moving an entity from one cluster to another or by swapping two clusters.
   - Stop searching when the score no longer improves.

Eight structural form types—partition, chain, order, ring, tree, hierarchy, grid, and cylinder—are used as candidate forms. Figure 2.3 illustrates these steps visually.

Summary

By applying an algorithm for discovering structural form to the US patent database, this research intends to leverage this large repository of existing design solutions to generate analogical inspiration to engineering designers. The methodology discovers structure in design repository databases, toward the ultimate goal of stimulating designers through design-by-analogy. We combine LSA with a Bayesian model for discovering structural form in data to gain useful insights into the nature of the design space. These results may provide a basis for automated discovery of cross-domain analogy in the form of a computational design stimulation tool.

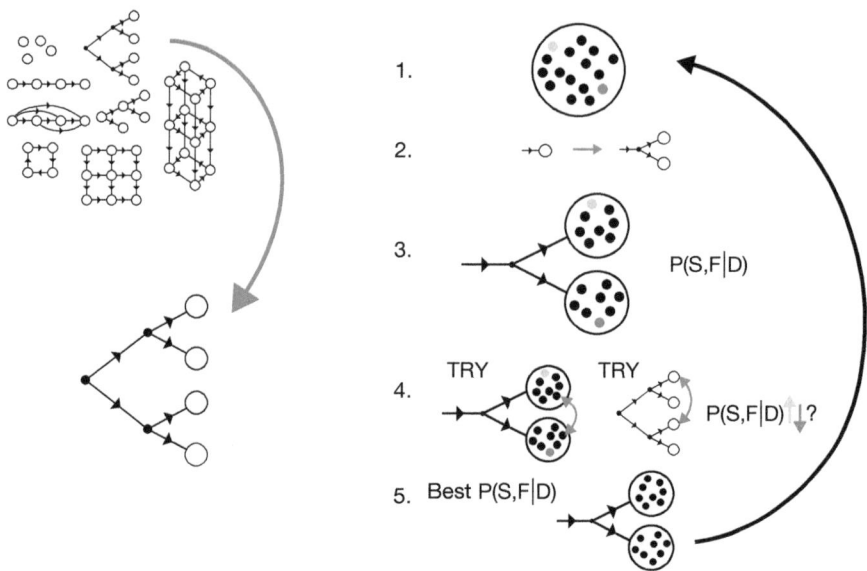

Figure 2.3 Pictorial illustration of steps of algorithm for discovering structural form. SOURCE: Reprinted with permission from *ASME Journal of Mechanical Design*, originally appearing in Fu et al. (2013).

A SWEET SPOT FOR DISTANCE OF ANALOGY

We next conducted a study to validate the structuring of patents generated using LSA and the Kemp and Tenenbaum's (2008) algorithm (Fu et al., 2013). In particular, it provided a new examination of the effect of "near" and "far" external analogical stimuli on design output quality. The "near" and "far" analogical stimuli for this study were chosen based on a structure of patents, created using the method discussed previously, resulting in clusters of patents connected by their relative similarity.

Choosing Initial Patent Set

A random number generator was used to create a list of random patent numbers, from which a subset of 45 patents were chosen that were classified within the US patent classification system as "Body Treatment and Care," "Heating and Cooling," "Material Handling and Treatment," "Mechanical

Manufacturing," "Mechanical Power," "Static," and "Related Arts." The full text of these 45 random "mechanical" patents was used to generate the structure to choose the analogical stimuli for the cognitive experiment.

Choosing Varying Stimulus Set

The best-fitting structure generated via the algorithm was a hierarchy and was used to choose five patents that were "near" and five patents that were "far" from the design problem description in the structure. The location of the design problem description was determined by calculating its semantic similarity to each cluster in the structure and choosing the node with the highest similarity. Near patents were chosen from nodes that were zero or one node away from the design problem. Far patents were chosen from nodes that were three nodes away from the design problem description. The five near and five far patents served as the varying stimulus set (see Figure 2.6).

Participants

This study was performed with 72 engineering students enrolled at a US university. All participants had adequate domain knowledge of engineering, and all but 2 had at least some design experience consisting of some combination of course-related design projects, industry experience, and structured design courses and training in design tools. There were 24 participants in each of the conditions.

Conditions

The independent variable was the patent distance in the structure as measured from design problem description position. There were three conditions:

1. "Near" patents—The varying stimulus set included five near patents, with each participant exposed to three of the five patents.

2. "Far" patents—The varying stimulus set included five far patents, with each participant exposed to three of the five patents.
3. Control—Participants received no external stimulus beyond the design problem.

COGNITIVE STUDY PROCEDURE

There were three phases in the study, similar to the previous study. Phase A was a pre-stimuli ideation phase, in which participants in all conditions worked for 10 minutes on the same energy from human motion design problem described in the previous study. In phase B, participants in the "near" and "far" conditions had 5 minutes to read and understand the three patents from their stimulus set; control condition participants continued to ideate during phase B. In phase C, all participants returned to ideating on the given design problem for 15 final minutes.

Results and Discussion

Metrics used to evaluate the design outcomes were the same as those described in the study presented previously in this chapter. The results from this study are presented together with the results of the previous study because the results were surprisingly in opposition. The previous study indicated that far-field analogical stimuli were associated with significantly *better* ideation performance over the near-field analogies in terms of novelty and quality. The results from the current study indicate that the patents designated as "far" patents were significantly less helpful to designers during their ideation than the patents designated as "near" in terms of their effect on novelty and quality of design output. Specifically, Figures 2.4 and 2.5, displaying the novelty and quality results from the previous and current studies, suggest what appear to be opposing conclusions regarding whether "near" or "far" patents are less harmful to the overall quality of design output.

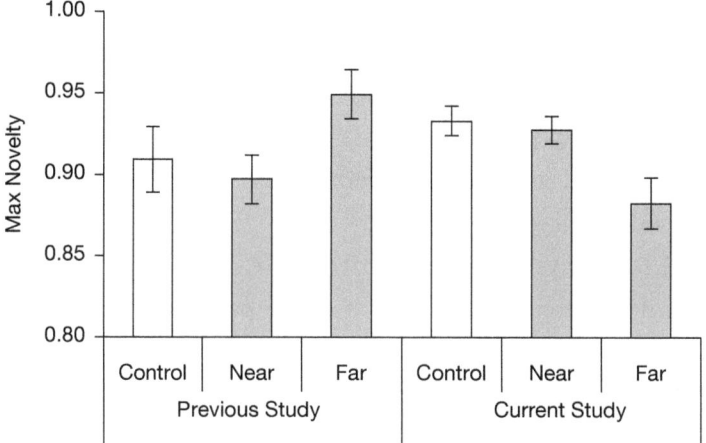

Figure 2.4 Effect of distance of analogical stimuli on novelty, previous study and current study results.
SOURCE: Reprinted with permission from *ASME Journal of Mechanical Design*, originally appearing in Fu et al. (2013).

These seemingly opposing results raised the following question: How consistently are the labels "near" and "far" used, here and in the research literature more broadly? "Near" and "far" often mean something different to each researcher and to each individual study or discussion. This

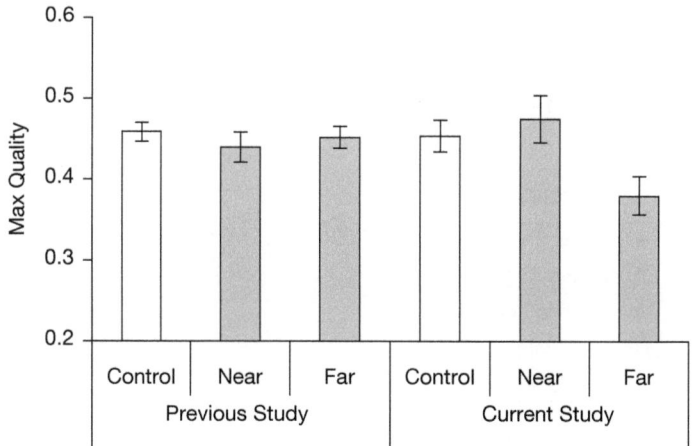

Figure 2.5 Effect of distance of analogical stimuli on quality, previous study and current study results.
SOURCE: Reprinted with permission from *ASME Journal of Mechanical Design*, originally appearing in Fu et al. (2013).

variation in definitions makes generalization about the effects of distance of analogy on design and ideation a difficult task. To reconceptualize distance more broadly and to reconcile the findings of the two studies, the same computational process used to choose the stimuli in this study was taken advantage of in order to understand how the definitions of "near" and "far" compared to one another.

Figure 2.6 displays the hierarchy structure created and used to choose the near and far patents for this study. The 8 patents, 4 "near" and 4 "far" from the previous study, were added to the structure of 45 random patents

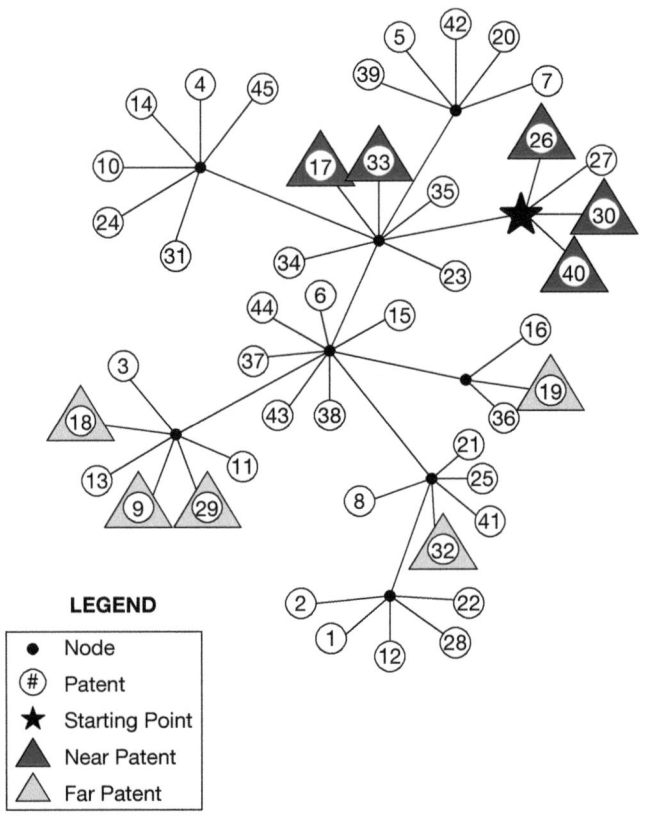

Figure 2.6 Original structure of 45 random patents used to choose stimuli sets in the current study.
SOURCE: Reprinted with permission from *ASME Journal of Mechanical Design*, originally appearing in Fu et al. (2013).

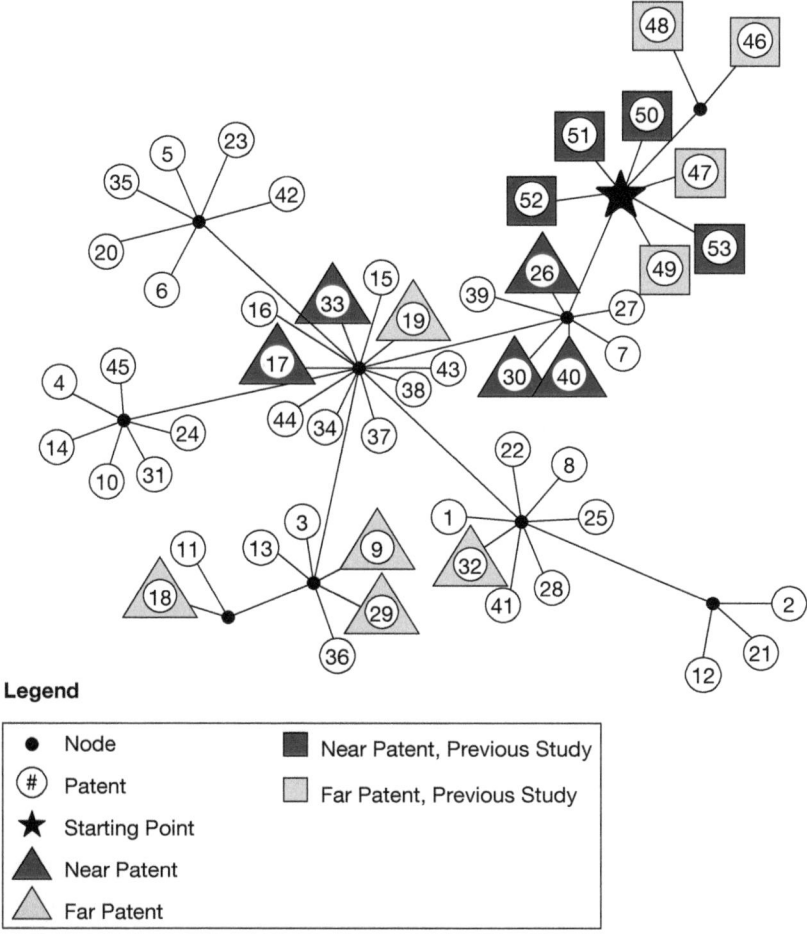

Figure 2.7 Original structure of 45 random patents with 8 patents from previous work. SOURCE: Reprinted with permission from *ASME Journal of Mechanical Design*, originally appearing in Fu et al. (2013).

used to choose stimuli for the current study. The resulting hierarchy structure is shown in Figure 2.7. Patents 46–49 and 50–53 were respectively "far" and "near" patents in the previous study. The structure shown in Figure 2.7 "quantifies" the relative distance of the external analogical stimuli to the design problem across the two different studies. As can be seen in the figure, the hand-picked near patents from the previous study are the closest set of patents to the design problem in the structure. Most saliently, the hand-picked far patents from the previous study are still closer than the near patents chosen by the algorithm within the randomly selected 45 patents. It is not surprising that the randomly selected 45 patents were generally far away and thus likely of lower use. As the effect of adding the hand-selected patents to the analysis shows, the relative distances in number of nodes in each distinct structure should be viewed not as an absolute measure but as a means for qualitative comparison.

These analytic results support the argument that near and far can have distinctly different meanings across the literature. In addition, it is a validation of the structuring methodology and its ability to portray relative analogical distance. This analysis method can be taken one step further by adding another 100 random patents to the space to see how the relationships might change in an even larger context pool, better mimicking how these patents might be situated in the entire patent database. The hierarchy structure in Figure 2.8 shows similarities to that in Figure 2.7 but with some interesting differences. The 8 patents from the previous work remain closely clustered around the design problem starting point. However, the patents from the current study are scattered in the structure in a more unexpected way; in many cases, the patents from this work called "near" are found in the same cluster as those called "far" or are found the same distance away from the starting point. That is, relative distance appears to be an emergent property that depends on the mixture of analogs included in the analysis.

Most important, these results suggest that there is a "sweet spot" for distance from the design problem when choosing analogical stimuli—in other words, there may be such a thing as "too near" and "too far" when searching for analogies to employ in design-by-analogy ideation practice. The "near" patents from the previous study appeared to be *too* "near" to be beneficial to designers as analogical stimuli. The "far" patents from the current study appeared to be *too* "far" to be useful to the designers. The results confirm that "near" and "far" are relative terms and depend on the characteristics of the potential stimuli. Furthermore, although the literature has shown that "far" analogical stimuli are more likely to generate innovative solutions with novel characteristics, there is such a thing as

Open Innovation Through Strategic Design-by-Analogy 37

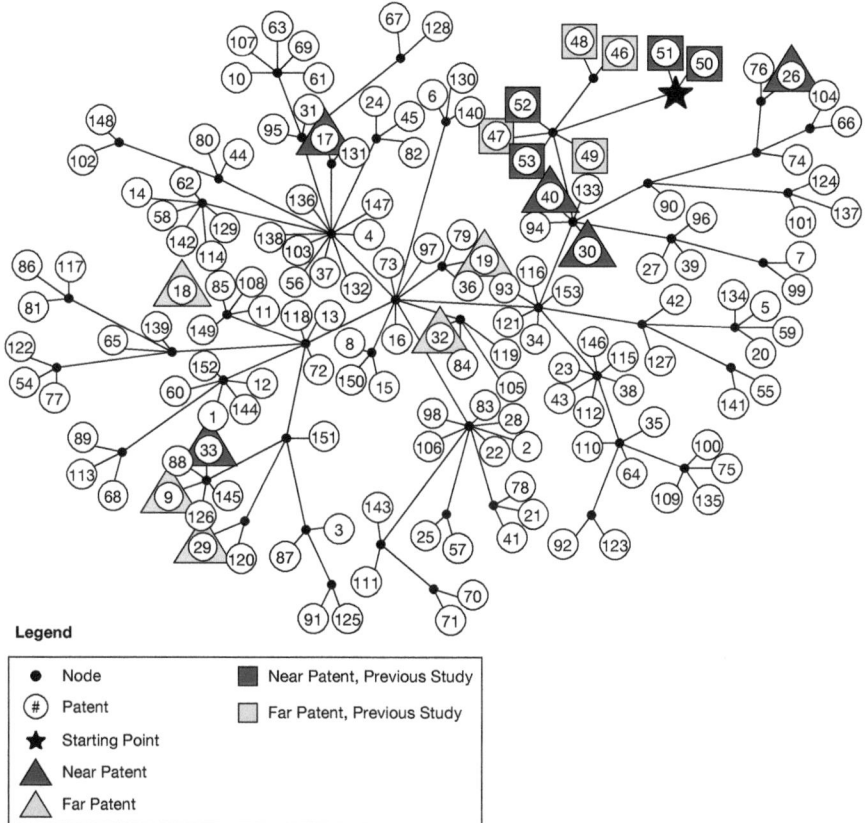

Figure 2.8 Original structure of 45 random patents with 8 patents from previous study and 100 additional random patents.
SOURCE: Reprinted with permission from *ASME Journal of Mechanical Design*, originally appearing in Fu et al. (2013).

too far. That is, if the stimuli are too distant, they then become harmful to the design process. Importantly, as well, the data mapping approach to identify analogies works, and it is able to impact the effectiveness of the design process.

CONCLUSIONS

The results presented here are important for several reasons. First, there is promise for the use of this structuring technique as the basis of a design inspiration tool for automatically finding design analogies. Designers often employ design-by-analogy by creating the analogies themselves, a stroke of luck, low hanging fruit, or genius. Currently, the most widely used method for searching for analogical inspiration in the patent database is the key word search. The results of key word searches can be an overwhelming undertaking to explore for design inspiration. There are also computational "innovation support tools" for sale to businesses and innovators (Goldfire, 2012). All of these methods and tools place the onus largely on designers to generate the terms or analogies of their own accord and comb through search results. The psychology literature has shown that the retrieval of far-field analogies is cognitively difficult (Gick & Holyoak, 1980). In addition, remindings tend to be constrained by surface similarity (Forbus et al., 1994), meaning the probability of retrieving surface dissimilar analogies is often low. Thus, a computational design tool that could find analogies in the "sweet spot" of moderate distance (as is recommended from the work presented here), which would not easily be located by a designer due to surface dissimilarity or rarity of occurrence, could be helpful in facilitating the practical use of the design-by-analogy method. With a large population of patents to build the structures, designers could have fast and relatively easy access to relevant analogies that could be useful or inspirational to them and that they may not have otherwise thought of or been able to find.

References

Adamson, R. E. (1952). Functional fixedness as related to problem solving: A repetition of three experiments. *Journal of Experimental Psychology, 44,* 288–291.

Anderson, J. R., & Schunn, C. D. (2000). Implications of the ACT-R learning theory: No magic bullets. In R. Glaser (Ed.), *Advances in instructional psychology*. Mahwah, NJ: Erlbaum.

Blanchette, I., & Dunbar, K. (2000). How analogies are generated: The roles of structural and superficial similarity. *Memory & Cognition, 28,* 108–124.

Boden, M. A. (2004). *The creative mind: Myths and mechanisms*. New York, NY: Routledge.

Casakin, H., & Goldschmidt, G. (1999). Expertise and the use of visual analogy: Implications for design education. *Design Studies, 20,* 153–175.

Chakrabarti, A. K., Dror, I., & Nopphdol, E. (1993). Interorganizational transfer of knowledge: An analysis of patent citations of a defense firm. *IEEE Transactions on Engineering Management, 40,* 91–94.

Chan, J., Fu, K., Schunn, C., Cagan, J., Wood, K., & Kotovsky, K. (2011). On the benefits and pitfalls of analogies for innovative design: Ideation performance based on analogical distance, commonness, and modality of examples. *Journal of Mechanical Design, 133*, 081004.

Chase, W. G., & Simon, H. A. (1973). The mind's eye in chess. In W. G. Chase (Ed.), *Visual information processing*. New York, NY: Academic Press.

Cheong, H., Shu, L. H., Stone, R., & Wood, K. L. (2008). *Translating terms of the functional basis into biologically meaningful keywords*. Paper presented at the Proceedings of the ASME International Design Engineering Technical Conference, New York, NY.

Chi, M. T. H., Feltovich, P. J., & Glaser, R. (1981). Categorization and representation of physics problems by experts and novices. *Cognitive Science, 5*, 121–152.

Chi, M. T. H., & Koeske, R. D. (1983). Network representation of a child's dinosaur knowledge. *Developmental Psychology, 19*, 29–39.

Christensen, B. T., & Schunn, C. D. (2005). Spontaneous access and analogical incubation effects. *Creativity Research Journal, 17*, 207–220.

Christensen, B. T., & Schunn, C. D. (2007). The relationship of analogical distance to analogical function and preinventive structure: The case of engineering design. *Memory & Cognition, 35*, 29–38.

Chrysikou, E. G., & Weisberg, R. W. (2005). Following the wrong footsteps: Fixation effects of pictorial examples in a design problem solving task. *Journal of Experimental Psychology: Learning, Memory & Cognition, 31*, 1134–1148.

Dahl, D. W., & Moreau, P. (2002). The influence and value of analogical thinking during new product ideation. *Journal of Marketing Research, 39*, 47–60.

Deerwester, S., Dumais, S. T., Furnas, G. W., & Landauer, T. K. (1990). Indexing by latent semantic analysis. *Journal of the American Society for Information Science, 41*, 391–407.

Dunbar, K., (1997). How scientists think: On-line creativity and conceptual change in science. In T. B. Ward, S. M. Smith, & J. Vaid (Eds.), *Creative thought: An investigation of conceptual structures and processes* Washington, DC: American Psychological Association.

Duncker, K. (1945). *On problem solving*. Washington, DC: American Psychological Association.

Dyer, J. H., Gregersen, H. B., & Christensen, C. M. (2011). *The innovator's DNA: Mastering the five skills of disruptive innovators*. Boston, MA: Harvard Business Review Press.

Foltz, P. W., Kintsch, W., & Landauer, T. K. (1998). The measurement of textual coherence with latent semantic analysis. *Discourse Processes, 25*, 285–307.

Forbus, K. D., Gentner, D., & Law, K. (1994). MAC/FAC: A model of similarity-based retrieval. *Cognitive Science, 19*, 141–205.

Fu, K., Chan, J., Cagan, J., Kotovsky, K., Schunn, C., & Wood, K. (2013). The meaning of "near" and "far": The impact of structuring design databases and the effect of distance of analogy on design output. *ASME Journal of Mechanical Design, 135*, 021007.

Gentner, D., & Markman, A. B. (1997). Structure mapping in analogy and similarity. *American Psychologist, 52*, 45–56.

Gick, M. L., & Holyoak, K. J. (1980). Analogical problem solving. *Cognitive Psychology*, *12*, 306–355.

Goldfire, I. M. (2012, February 19). *Invention machine Goldfire: Unleashing the power of research.* Available at http://inventionmachine.com/products-and-services/innovation-software/goldfire-Research.

Goldschmidt, G., & Smolkov, M. (2006). Variances in the impact of visual stimuli on design problem solving performance. *Design Studies, 27*(5), 549–569.

Green, M., Dutson, A., Wood, K. L., Stone, R., & McAdams, D. (2002). *Integrating service-oriented design projects in the engineering curriculum.* Paper presented at the proceedings of the 2002 American Society for Engineering Education Annual Conference and Exposition.

Green, M., & Wood, K. L. (2004). Service-learning approaches to international humanitarian design projects: Assessment of spiritual impact. In *Proceedings of the 2004 Christian Engineering Education Conference.*

Griffiths, T. L., Kemp, C., & Tenenbaum, J. B. (2008). Bayesian models of cognition. In R. Sun (Ed.), *Cambridge handbook of computational psychology* (pp. 59–100). New York, NY: Cambridge University Press.

Indukuri, K. V., Ambekar, A. A., & Sureka, A. (2007). *Similarity analysis of patent claims using natural language processing techniques.* Paper presented at the International Conference on Computational Intelligence and Multimedia Applications.

Jansson, D. G., & Smith, S. M. (1991). Design fixation. *Design Studies, 12*, 3–11.

Kaplan, C., & Simon, H. A. (1990). In search of insight. *Cognitive Psychology, 22*, 374–419.

Kemp, C., & Tenenbaum, J. B. (2008). The discovery of structural form. *Proceedings of the National Academy of Sciences of the USA, 105*, 10687–10692.

Knoblich, G., Ohlsson, S., Haider, H., & Rhenius, D. (1999). Constraint relaxation and chunk decomposition in insight problem solving. *Journal of Experimental Psychology: Learning, Memory, and Cognition, 25*, 1534–1555.

Landauer, T. K., Foltz, P. W., & Laham, D. (1998). An introduction to latent semantic analysis. *Discourse Processes, 25*, 259–284.

Linsey, J., Murphy, J., Markman, A., Wood, K. L., & Kortoglu, T. (2006). *Representing analogies: Increasing the probability of innovation.* Paper presented at the ASME International Design Theory and Method Conference, Philadelphia, PA.

Linsey, J. S., Wood, K. L., & Markman, A. B. (2008). Modality and representation in analogy. *Artificial Intelligence for Engineering Design, Analysis & Manufacturing, 22*, 85–100.

Maier, N. R. F. (1931). Reasoning in humans: II. The solution of a problem and its appearance in consciousness. *Journal of Comparative Psychology, 12*, 181–194.

Markman, A. B., & Wood, K. L. (2009). *Tools for innovation: The science behind practical methods that drive new ideas.* New York, NY: Oxford University Press.

Marsh, R. L., Ward, T. B., & Landau, J. D. (1999). The inadvertent use of prior knowledge in a generative cognitive task. *Memory & Cognition, 27*, 94–105.

McKoy, F. L., Vargas-Hernandez, N., Summers, J. D., & Shah, J. J. (2001). *Influence of design representation on effectiveness of idea generation.* Paper presented at the ASME IDETC and CIE, Pittsburgh, PA.

Moss, J., Cagan, J., & Kotovsky, K. (2007). *Design ideas and impasses: The role of open goals*. Paper presented at the Proceedings of the 16th International Conference on Engineering Design.

Moss, J., Kotovsky, K., & Cagan, J. (2007). The influence of open goals in the acquisition of problem relevant information. *Journal of Experimental Psychology: Learning, Memory, and Cognition, 33*, 876–891.

Newell, A. (1990). *Unified theories of cognition*. Cambridge, MA: Harvard University Press.

Ohlsson, S. (1992). Information-processing explanations of insight and related phenomena. In K. Keane (Ed.), *Advances in the psychology of thinking*. Hertfordshire, UK: Harvester Wheatsheaf.

Purcell, A. T., & Gero, J. S. (1992). Effects of examples on the results of a design activity. *Knowledge-Based Systems, 5*, 82–91.

Purcell, A. T., & Gero, J. S. (1996). Design and other types of fixation. *Design Studies, 17*(4), 363–383.

Rantanen, K., & Domb, E. (2002). *Simplified TRIZ: New problem solving applications for engineers and manufacturing professionals*. Boca Raton, FL: CRC Press.

Shah, J. J., Vargas-Hernandez, N., & Smith, S. M. (2003). Metrics for measuring ideation effectiveness. *Design Studies, 24*, 111–134.

Smith, S. M., & Blankenship, S. E. (1991). Incubation and the persistence of fixation in problem solving. *American Journal of Psychology, 104*, 61–87.

Smith, S. M., Ward, T. B., & Schumacher, J. S. (1993). Constraining effects of examples in a creative generation task. *Memory & Cognition, 21*, 837–845.

Stone, R., & Wood, K. L. (2000). Development of a functional basis for design. *Journal of Mechanical Design, 122*(4), 359–370.

Szykman, S. Sriram, R. D., Bochenek, C., Racz, J. W., & Senfaute, J. (2000). Design repositories: Engineering design's new knowledge base. *IEEE Intelligent Systems*, 48–55.

Terwiesch, C., & Ulrich, K. T. (2009). *Innovation tournaments*. Cambridge, MA: Harvard Business School.

Tseng, I., Moss, J., Cagan, J., & Kotovsky, K. (2008). The role of timing and analogical similarity in the stimulation of idea generation in design. *Design Studies, 29*, 203–221.

Weisberg, R. W. (2009). On "out-of-the-box" thinking in creativity. In K. W. A. Markman (Ed.), *Tools for innovation* (pp. 23–47). New York, NY: Oxford University Press.

White, C., & Wood, K. L. (2010). *Influences and interests in humanitarian engineering*. Paper presented at the proceedings of the 2010 ASEE annual conference, Global Colloquium on Engineering Education, 2010.

Wilson, J. O., Rosen, D., Nelson, B. A., & Yen, J. (2010). The effects of biological examples in idea generation. *Design Studies, 31*, 169–186.

Zhang, R., Cha, J., & Lu, Y. (2007). *A conceptual design model using axiomatic design, functional basis and TRIZ*. Paper presented at the Proceedings of the 2007 IEEE.

3

Getting the Most out of Brainstorming Groups

PAUL B. PAULUS, JUBILEE DICKSON, RUNA KORDE,
RAVIT COHEN-MEITAR, AND ABRAHAM CARMELI ∎

Most corporations indicate that innovation is one of their top priorities. For example, according to the McKinsey Global Survey, approximately 70% of corporate leaders from a broad range of industries indicated that innovation is among their top three priorities for driving growth (Barsh, Capozzi, & Mendonca, 2007). In a survey conducted by Bain & Company among executives throughout the world, two-thirds indicated that their companies, which have more than $100 million in revenue, made innovation one of their top three priorities. However, it was also found that fewer than one-fourth believed that their companies were effective innovators, and just one in five noted that their companies were strong at "breakthrough" innovation (Almquist, Leiman, Rigby, & Roth, 2013a).

One crucial problem concerns the reactions toward innovative ideas. Scholars have pointed out that innovative ideas tend to receive negative reactions (Mueller, Melwani, & Goncalo, 2012), and thus the innovation process in which people come up with new ideas, champion their ideas, and seek support for them may involve significant hurdles. Furthermore, given the increasing complexity of technical and social problems, organizations allocate tasks to work groups, acknowledging that collaboration of diverse experts is a key for driving innovative solutions in the marketplace. This collaboration process can involve a variety of forms—face-to-face meetings, virtual or computer-mediated interactions, or a mixture of these two forms. Following a line of research on the effectiveness of the collaborative innovative process (Paulus & Coskun, 2012), we focus on group brainstorming and discuss the benefits and drawbacks of exchanging ideas in collaborative settings. This extensive literature sheds light on hurdles associated with the creative group process and the ways to optimize it (De Dreu, Nijstad, Bechtoldt, & Baas, 2011; Paulus & Coskun, 2012). However, there has been little controlled research in actual work or organizational settings. Also, practitioners often find it difficult to effectively apply the research findings to the workplace. Much of the controlled research on brainstorming has used ad-hoc groups that brainstorm for short periods of time under specific conditions. However, in real-world settings, idea generation does not occur under such constrained conditions. In this chapter, we present some new knowledge from a work setting and discuss the applicability and extension of the research findings to typical work environments.

CREATIVE GROUP PROCESSES

Problems with "Natural Meetings"

In most organizations, a considerable amount of time is spent in meetings to discuss plans, problems, and potential solutions. Although such meetings typically serve a useful purpose for sharing information and decision-making, face-to-face meetings bear some shortcomings and may not be an optimal format for coming up with innovative ideas. We summarize these shortcomings in Box 3.1 and discuss them in further detail here. One difficulty is that only one person can talk at one time. This confines full sharing of ideas in a limited time. A key finding in research on group brainstorming is that group members who verbally share ideas will

Box 3.1

KEY PROCESSES AND TASK FACTORS IN BRAINSTORMING

Inhibiting Processes
  Production blocking
  Concern about evaluation
  Premature fixation
  Focus on consensus
  Failure to tap expertise diversity
  Deferment to high-status members

Facilitating Processes
  Attention
  Motivation
  Social comparison
  Cognitive stimulation
  Building on shared ideas
  Functional or expertise diversity

Facilitating Task Factors
  Minimize production blocking (electronic brainstorming brainwriting, small groups)
  Minimize concern with evaluation
  Task structure (one issue at a time, prior consideration of categories)
  Brainstorming rules (focus on quantity)
  Brief breaks
  Performance feedback/comparison
  Alternating alone and group ideation sessions
  Prosocial personality characteristics
  Positive attitude to groups and diversity
  Facilitator/leader guidance and support
  Training

generate fewer ideas than a similar group of individuals generating ideas on their own (Diehl & Stroebe, 1987). This discrepancy increases with the increased size of the group—so the larger the group, the larger the potential loss of innovative ideas (Bouchard & Hare, 1970). This problem has been referred to as *production blocking* because whenever one group member talks, others cannot present their ideas.

Another difficulty concerns the timing in which ideas are presented; that is, ideas presented early in the process tend to have more impact and may lead to premature fixation on those ideas (Kohn & Smith, 2011). This is unfortunate because the initial ideas tend to be more easily accessible and "common," whereas more original ideas tend to be generated later in the discussion process because at that point group members will be tapping ideas or categories that are relatively low in accessibility and therefore more unique (Brown & Paulus, 2002; Paulus, Kohn, Arditti, & Korde, 2013). There is also a tendency for groups to focus on agreement rather than diversity of perspectives (Stasser, Abele, & Parsons, 2012), and there may be a premature consensus on ideas that are not particularly innovative. Groups also vary in the status and expertise level of the group members. Thus, there may be a tendency for participants to defer to higher status or expert members (van der Vegt, Bunderson, & Oosterhof, 2006). However, research suggests that it is important for low expertise or new members and members who have unique perspectives to share them with the group (Choi & Thompson, 2005; Nemeth & Nemeth-Brown, 2003). Thus, despite the importance of face-to-face meetings in organizational functioning, other forms of interaction and procedures are essential to fully realize the innovative potential of groups in organizations.

Theoretical Basis for Enhanced Creativity in Groups and Teams

Most of us intuitively believe that collaboration is beneficial for innovation (Paulus, Larey, & Ortega, 1995). There are in fact a number of theoretical models that have been proposed that provide a basis for such an expectation (Nijstad & Stroebe, 2006; Paulus & Brown, 2003, 2007). One basis is *motivational*. It is important that group members have external and internal motivation for creativity (De Dreu et al., 2011). For example, productive group members may stimulate others to a higher level of productivity (Paulus & Dzindolet, 1993), and group leaders can motivate group members to exchange knowledge that, in turn, facilitates creative problem solving (Carmeli, Gelbard, & Reiter-Palmon, 2013), engagement in search and discovery (Carmeli & Paulus, 2015), and innovation performance (Collins & Smith, 2006).

The other basis is *cognitive*. Exposure to the ideas of others can potentially stimulate group members to think of ideas or categories of ideas they would not have otherwise considered. This can increase the number of ideas generated and the potential for more original ideas (Dugosh & Paulus, 2005; Hargadon, 2008), or a combination of knowledge that has been exchanged, which is a key to innovation (Carmeli & Azerual, 2009). However, how the interaction and the exchange processes are structured makes a significant difference in the extent to which the cognitive potential is realized. It is important for group members to carefully attend to the shared ideas so that they can yield additional associations (Brown & Paulus, 2002; Michinov, Janet, Métayer, & Le Hénaf, 2015). When the interaction is structured so that participants more fully tap the various categories of a problem, more ideas and more unique ideas are generated (Deuja, Kohn, Paulus, & Korde, 2014; Rietzschel, Nijstad, & Stroebe, 2007). Exposure to diverse perspectives in groups can increase the number and quality of ideas generated (Paulus & van der Zee, 2015). Furthermore, group members should be motivated to process and build on the shared ideas (De Dreu et al., 2011). These issues are discussed in more detail next. The basic processes and task factors that are important in group brainstorming are summarized in Box 3.1.

Enhancing the Number of Ideas for Both Groups and Individuals

Several interventions may facilitate the ideation process among both groups and individuals in a relatively similar manner. These interventions either provide increased motivation or enhance the cognitive ideation process. Since brainstorming was promoted as an effective tool for increasing innovation (Osborn, 1957), there has been an emphasis on using the rules or guidelines that were originally emphasized by Osborn. He argued that it is important for ideas to be presented without immediate evaluation or feedback. Critical feedback was deemed to inhibit the free flow of ideas because individuals would tend to personally censor ideas they thought might not yield positive reactions. Thus, they would not express whatever came to their minds—another of Osborn's rules. Some researchers suggest that a concern about evaluation of ideas inhibits the number of ideas generated (Camacho & Paulus, 1995; Diehl & Stroebe, 1987), whereas others have argued that allowing critical evaluation may not hinder the generation process (Nemeth & Ormiston, 2007). We suggest that currently the case is stronger for the *minimal evaluation approach*, with evaluation being reserved for later phases of the innovation process.

Osborn (1957) proposed that groups should focus on generating a high number of ideas (*quantity goal*) because this would increase the likelihood of coming up with more good ideas. This perspective is consistent with other evidence in science that creative genius is often associated with high levels of activity (Simonton, 2004). A study that examined this rule (Paulus, Kohn, & Arditti, 2011) found that having individual students focus on generating a high number of ideas was indeed more beneficial than having them focus on generating high-quality ideas; a focus on quantity increased both the number of ideas generated and the number of good ideas (original and feasible). In another study, it was found that providing students with high quantity goals increased the number of ideas they generated (Paulus & Dzindolet, 1993).

A fourth Osborn rule is that group members should *build on the ideas of others*. Although this seems a sensible and natural guideline, it has been examined in only one study (Kohn, Paulus, & Choi, 2011). In this study, students were asked to build on ideas that others had generated either alone or in a group. It was found that when participants were presented with unique ideas in comparison to relatively common ideas, the groups, but not individuals, were able to generate more novel and feasible combinations. Thus, groups may be uniquely beneficial when group members are considering ways to build on very novel ideas. Periodic presentation of common ideas during brainstorming seems to stimulate more creative ideas for individuals than unique ideas (Dugosh & Paulus, 2005). A very unique idea may have a lower probability of overlapping with the associational network of a particular individual. However, in a group collaborative context, there is an increased probability that one of the group members can relate to the unique idea. This can then aid the other group members in developing novel combinations with this unique idea.

Because coming up with new ideas appears to be a generally difficult task, it is important to provide some *external motivation* for brainstormers. Providing them with high goals or reference points is one way to increase their motivation and the generation of a large number of ideas (Larey & Paulus, 1995; Paulus & Dzindolet, 1993). Providing group members with feedback that other groups had generated more ideas in a previous session also can increase motivation to generate more ideas in subsequent sessions (Coskun, 2000). In fact, just having a chance to compare one's rate of generating ideas with others can motivate increased performance (Michinov et al., 2015; Munkes & Diehl, 2003; Paulus, Larey, Putman, Leggett, & Roland, 1996), possibly because this increases feelings of competition.

Even over short periods of time, individuals and groups typically exhibit a fairly rapid decline in the rate of ideas generated. One way to counteract that decline is to provide *brief breaks* during the brainstorming session (Kohn & Smith, 2011; Paulus, Nakui, Putman, & Brown, 2006). Such breaks not only allow individuals some time for cognitive rest but also may reduce fixation on a limited set of ideas. Breaks can serve as an incubation opportunity. They allow individuals to reflect on the ideas already generated (without the competing pressure of continuous idea generation) and may enable them to come up with some new directions for subsequent sessions. Currently, there is not enough evidence to suggest how often breaks should be provided, how long the breaks should be, or what activities should occur during breaks. The perspective by Smith (2003) is that a "change of scenery" might be most helpful to break out of a fixated train of thought. Environmental research also suggests the benefit of breaks in natural environments (Berman, Jonides, & Kaplan, 2008), but this has not been examined for creativity tasks.

In addition to providing brief breaks, it may be helpful to provide some *task structure*. When people are provided with a broad topic, they may not fully tap the various aspects of that issue. Providing a narrower topic for ideation can increase the extent to which individuals generate more unique ideas on that issue (Rietzschel et al., 2007; Rietzschel, Nijstad, & Stroebe, 2014). Focusing on one aspect of a problem at a time can also increase the number of ideas generated as compared to considering multiple aspects of the problem at one time (Coskun, Paulus, Brown, & Sherwood, 2000; Dennis, Valacich, Connolly, & Wynne, 1996). It is also beneficial to have individuals first think about the various aspects of the problem before beginning a brainstorming session (Deuja et al., 2014). These latter two techniques can prevent the tendency to focus only on a limited number of categories of a problem and to stop generating ideas before a broad range of alternative categories have been explored.

When groups are sharing ideas, group members may provide stimulation by generating ideas in different categories. This can be particularly stimulating if the categories shared are not closely related semantically (Baruah & Paulus, 2011). This supports the idea that some degree of intellectual diversity in groups can be beneficial for creativity. However, Baruah and Paulus also found that it was beneficial if participants were all focusing on generating ideas in the same set of categories rather than all focusing on different categories at the same time. Having a common focus helped increase the flow of similar ideas, which in turn enhanced the total number of ideas generated. Thus, participants in diverse groups should share their diverse and semantically unrelated categories of knowledge, but these should be shared in sequence so that the group members can have a shared focus on one topic at a time.

How to Enhance the Effectiveness of Group Brainstorming

It is apparent that one of the major drawbacks to brainstorming in meetings is production blocking. That is, the larger the group, the greater the blocking. Even with a group as small as four, the number of ideas generated is often 50% lower than that of nominal groups (Diehl & Stroebe, 1987). Thus, one key to effective group brainstorming is to keep the group as small as possible. Various reviews suggest that the "ideal group" should be dyadic because, at that level, there is typically a small degree of production loss relative to that of nominal groups. However, it may be necessary to include more people in the group who represent different areas of expertise. For example, a review of the team innovation literature indicates that larger teams are rated as more innovative (Hülsheger, Anderson, & Salgado, 2009). That is not surprising because it may simply reflect the "additive effect" of having more team members with possibly a greater diversity of perspectives. However, in interacting settings, increased group size reduces the opportunity of group members to contribute. In that case, it is suggested to keep the group as small as possible but still large enough to represent the relevant diverse perspectives.

One effective way to overcome production blocking in groups is to use *interaction modalities* that do not require waiting on others before making one's own contributions. This can be done easily with various computer-based platforms or group-decision systems. Studies have shown that when groups interact by means of such systems, much of the production loss of group interaction disappears (Dennis & Williams, 2003). Reviews of the literature on such electronic brainstorming indicate that there may actually be a benefit to larger groups. With groups of nine or more members, electronic group brainstorming can lead to more ideas than generated by similar-sized nominal groups. For smaller groups, there may still be a small deficit (De Rosa, Smith, & Hantula, 2007). Hence, even in small electronic groups, attending to the ideas of others, while brainstorming, may slow the ideation process. In larger groups, the large number of ideas generated may provide enough stimulation of additional ideas to overcome any negative or distracting effects of monitoring the ideas of others. However, Paulus et al. (2013) noted that even in large groups, the benefit of group interaction was only a few additional ideas.

Brainwriting

Another technique that enables groups to overcome production blocking is brainwriting (Heslin, 2009). This technique has been utilized for many years, and one popular version is group brainwriting. Group brainwriting can be structured in a variety of ways, but it typically involves group members writing ideas on slips of paper and passing these on to the other group members for perusal. Ideas are read as participants complete writing their ideas on slips and pass them on. The advantage of this technique is that participants can generate ideas as they occur but also access ideas from others. Clearly, taking time to read ideas from others takes time from generating one's own ideas or may interrupt one's cognitive flow. However, one could assume that the stimulating benefits of shared ideas in this paradigm will outweigh the distraction effect, as in the case of large, electronic brainstorming groups. Only one study has compared electronic brainstorming and brainwriting. This study found that when the task demands are similar, the performance with these two methods is essentially the same (Michinov, 2012). The advantage of brainwriting is that it does not require an electronic support system. However, electronic systems have the advantage of generating a summary of the ideas shared.

There are several reasons why brainwriting might be more beneficial than electronic brainstorming for enhancing performance. In electronic brainstorming, there is typically no way to ensure or "force" participants to process the shared ideas. In brainwriting, participants are instructed to read the ideas as they are passed along in the group. Our observation of group brainwriting is that participants actually read the shared ideas (Kohn et al., 2011). Another reason why brainwriting might be more beneficial is that it may induce more feelings of competition among participants. As slips are passed from one member to another, there is pressure to keep generating ideas at a similar pace. Thus, unlike electronic brainstorming, there may not be a deficit with a small group of brainwriters relative to a nominal group of individual writers.

Unfortunately, there have been no systematic comparisons of brainwriting groups versus electronic brainstorming that allow for a determination of differential effects of group size. Nevertheless, several studies have compared brainwriting groups with nominal groups. Paulus and Yang (2000) found that brainwriting groups of four outperformed nominal groups of four by approximately 40% in a 15-minute session. In a subsequent 15-minute session, both groups continued to brainstorm individually. Those who had previously brainstormed as an interactive group outperformed the nominal groups by approximately 90%. One potential problem with the study by Paulus and Yang was that individuals generated ideas on one sheet of paper, whereas group members used individual slips of paper. Writing on one sheet of paper may have led individuals to have an inflated perception of their performance and thus lowered their motivation to continue to generate ideas at a high rate. Goldenberg, Larson, and Wiley (2013) found that when both interactive and nominal groups used individual slips, their overall performance did not differ. Nonetheless, the fact that groups were as good as nominal groups can still be considered a positive finding.

We were able to assess brainwriting at a major high-technology corporation (Paulus, Korde, Dickson, Carmeli, & Cohen-Meitar, 2015). We asked engineers to generate ideas for new technological innovations for their corporation on individual slips of paper. They did so either in groups of three or alone for 15 minutes. A comparison of the performance of these two sets of groups indicated that the group brainwriters generated 39% more ideas than the individual brainstormers. We had a limited sample size because only 57 employees were available for the study, so these results did not reach statistical significance. However, one can presume that a 39% increase in ideas is of importance for many organizations that seek to develop a competitive technological edge.

Alternating Group and Individual Brainstorming

Although experimental studies have focused on comparing the relative efficacy of individual and group brainstorming, in real-world contexts there is no strict segregation of these two modes of idea generation. In most innovation settings, generating novel ideas involves a mix of individual and group ideation. Individuals may think of ideas prior to a meeting and then share their ideas during the meeting. The discussion during the meeting may stimulate more ideas among the group members. After the meeting, additional ideas may occur to individuals as they continue to reflect on the shared ideas.

Although this may be a prototypical process in real-world settings, there have been only a few evaluations of the benefits of mixing alone and group brainstorming. The findings of these studies are mixed with regard to whether it is better to generate ideas alone before group discussion or to do group discussion prior to reflecting privately about the issue (Baruah & Paulus, 2009). Paulus and Yang (2000) found that when participants in brainwriting groups were provided a subsequent session to generate additional ideas, they continued to generate more ideas than those who had a prior solitary brainwriting session. Thus, there appeared to be a positive carryover of the associations generated during the group session.

We were able to compare the group-to-alone sequence with an alone-to-group sequence in the high-technology corporation (Paulus et al., 2015) and found that the group-to-alone sequence led to 63% more ideas and approximately twice as many good ideas (those high in both novelty and feasibility). The alone-to-group condition of course did not have a chance to build on the ideas shared as a group. If we had added a third session in which participants could reflect on the shared ideas, we probably would have observed a similar positive result. In fact, one could argue that the option to first generate ideas individually has the benefit of preventing fixation on a limited range of ideas that could occur in a group idea-sharing session (Ziegler, Diehl, & Zijlstra, 2000). Hence, the most optimal condition would be one that would alternate group and alone ideation (Korde, 2014). However, no published studies had compared variations in group and individual brainstorming with conditions in which only individual or group brainstorming occurred.

We were able to conduct such a study in the high-technology corporation (Paulus et al., 2015). Engineers generated ideas about market and product innovation in groups of three by writing ideas on slips of paper. In one condition, the engineers generated ideas as a group for 30 minutes. In another condition, they generated ideas alone for 8 minutes, shared ideas for 3 minutes, generated ideas alone again for 8 minutes, then shared ideas for 3 minutes, and, finally, generated alone again for 8 minutes. The condition in which the participants alternated individual and group brainwriting generated 36% more ideas and 26% more good ideas (both high in novelty and feasibility) than those who only generated ideas as a group. Thus, there is some empirical evidence to support the practice of varying solitary and group brainstorming even in a short-term session. However, further research is required to determine the consistency of this effect and the role of other factors such as time. What is the optimal mix of time for individual and group brainstorming? Is it best to have these sessions contiguous to one another, or should there be breaks in between to allow for incubation processes? Because associations that are stimulated in a group context may dissipate quickly over time, it may be important to have individual reflection/ideation sessions immediately after group sharing sessions. However, it will also be important to tap ideas that arise as one reflects on the shared ideas after some incubation time. Thus, the number and quality of the ideas being generated may depend on the exact procedure used. Linsey et al. (2011) compared various techniques for an engineering task. They found that a gallery method (writing ideas and sketches on sheets and then allowing group members to see them) enhanced the generation of high-quality ideas, and a rotational method (ideas are shared as they are being generated) using both words and sketches was useful for generating a large number of ideas. They concluded that an optimal procedure may be for individuals first to generate ideas using the gallery method and then switch to the rotational method for the further development of the ideas or for generation of additional ideas.

Experience, Facilitators, and Training

Generating ideas in groups is more complicated than doing so alone. One has to handle the multiple tasks of generating one's own ideas, listening to ideas from others, coordinating the sharing of ideas, and building on the ideas from others. The literature suggests that groups do not do this very well. However, the process can be cultivated significantly by a facilitator who helps manage the group process (Kramer, Fleming, & Mannis, 2001; Oxley, Dzindolet, & Paulus, 1996). This facilitator can ensure that all participants make a contribution, that they follow the rules, and that they persist on the task even when it becomes increasingly difficult. However, it is not desirable or feasible for every group to have a facilitator for creativity tasks.

Fortunately, it appears that even short training sessions can enhance the extent to which brainstorming groups can function more effectively. For example, Baruah and Paulus (2008) found that a 1-hour session that involved various tips, practice, and feedback increased the number and quality of ideas in groups. One factor that has not been examined is the effect of experience in brainstorming. However, other studies suggest that groups that have experience and training together will increase their performance level (Moreland, Argote, & Krishnan, 1996). Such training can enhance the shared mental model of the group (the collective understanding of how the group is supposed to function) and the group's transactive memory (the extent to which group members have an accurate understanding of the expertise diversity in the group) (Moreland et al., 1996). This appears to be particularly important for diverse groups. For example, Watson, Kumar, and Michaelson (1993) found that diverse groups of MBA students had inferior performance on creativity tasks early in the semester relative to homogeneous groups, but this difference disappeared toward the end of the semester.

Diversity

Previously, we mentioned the potential benefit of diversity of expertise or background for collaborative innovation. This makes obvious sense because many of today's problems are complex and require people from various disciplines or backgrounds to work together to solve. Reviews of the literature indicate that diversity along demographic, ethnic, and cultural dimensions may have a largely negative effect in teams (Bell, Villado, Lukasik, Belau, & Briggs, 2011; Mannix, Neale, & Goncalo, 2009; van Knippenberg & Schippers, 2007). Even research on functional diversity (variation in expertise) in teams generated inconclusive findings with regard to its contribution to innovation (Keller, 2001).

There are a variety of potential reasons for this mixed picture on the potential benefits of functional diversity. One is that the measures employed (mostly verbal reports) do not accurately reflect the real benefit of functional diversity (van Dijk, van Engen, & van Knippenberg, 2012). Alternatively, it could be that the difficulties in communication and comprehension among disciplines may reduce the potential beneficial impact of such diversity on innovation. Furthermore, functional diversity in groups should be beneficial to the extent that such diversity relates to the assigned task (Paulus & van der Zee, 2015). Research has thus far not examined the role of task relevance on the relationship between functional diversity and innovation. However, studies of group creativity that have examined cultural or demographic diversity have demonstrated positive effects of diversity (Paulus & van der Zee, 2015). This suggests that creativity tasks may be more likely than other types of tasks to benefit from diversity, possible because these types of tasks may more effectively tap differences in experience and perspectives. Furthermore, groups that consist of people who have a positive attitude toward diversity are more likely to benefit from cultural diversity in brainstorming sessions (Nakui, Paulus, & van der Zee, 2011).In addition, the team that has experience working as a team and training on best practices for team innovation may also enhance the innovative potential of diverse groups.

Individual Differences

Research on groups has tended to ignore the role of personal characteristics in part because the influence of group processes tends to overwhelm their impact. That is, the task structure and social context may be more influential than personal characteristics (Snyder & Ickes, 1985). Of the Big Five personality characteristics (agreeableness, conscientiousness, neuroticism, extraversion, and openness; Goldberg, 1992), openness appears to be most influential (Feist, 2010; Hoff, Carlsson, & Smith, 2011), with those who are more open to experience tending to be more creative and thus yielding potentially more overall creativity in groups. Whereas extraverts tend to be more verbose in individual brainstorming (Putman, 2001), introverts have been often found to be more creative (Hoff et al., 2011). However, it seems likely that a balance of introversion/extraversion may be optimal (Csikszentmihályi, 1997). Introverts may be able to "dig deeper" intellectually to develop novel ideas, but extraverts may be better at developing and promoting ideas (Hoff et al., 2011). Thus, a group or team that has both extraverts and introverts, or people who have a balance of extraversion and introversion, may be optimal for group creativity. In a team that consists of mostly extraverts, there may be much competition for speaking time in face-to-face meetings (Barry & Stewart, 1997). A process that involves both individual and group sessions would allow introverts to generate their ideas and build on previously shared ideas privately for later sharing with the group. Brainwriting or electronic procedures would be quite useful for this type of process.

Only a few studies have provided evidence of the impact of personality in group brainstorming. Not surprisingly, these have focused on the social characteristics. Generally, group members who like working in groups (Larey & Paulus, 1999) and are low in social interaction anxiety (Camacho & Paulus, 1995) perform better in groups. Group members who are high in interaction anxiety tend to perform poorly in brainstorming groups. Groups who are low in social interaction anxiety actually outperform nominal groups of low anxious individuals (Camacho & Paulus, 1995). Prosocial motivation (a focus on collective rather than personal gain) is also related to higher levels of performance in brainstorming groups (Bechtoldt, De Dreu, Nijstad, & Choi, 2010). This supports the motivated information processing theory of De Drue et al. (2011), which suggests that both external and internal motivation are important for creativity in groups, with the prosocial motivation being a critical personal factor.

Selection of Ideas

The brainstorming literature has focused primarily on how to increase the number of ideas generated by groups and individuals. This seems to be a reasonable goal because the number of good ideas is related to the number of ideas generated (Paulus et al., 2011). However, in most cases, organizations are interested in generating the best ideas—ones that are original but also feasible (doable and likely to be accepted by consumers). Unfortunately, quantity and average quality of ideas tend to be unrelated (Putman & Paulus, 2009). Thus, it becomes important to develop an effective selection mechanism for the highest quality ideas.

An organization may be fortunate to have some key people or leaders who have a "nose" for good ideas that will be successful. However, controlled studies have found that individuals and groups generally have a difficult time choosing the best ideas from a large group of ideas (Putman & Paulus, 2009; Rietzschel, Nijstad, & Stroebe, 2006). The average quality of the selected ideas tends not to be better than the overall average quality of all the ideas generated. Moreover, there tends to be a bias toward feasible ideas. Therefore, it is important to discover the best way to perform the idea selection process. Some research indicates that groups may do a better job than individuals (Larey & Paulus, 1999). Group members may be able to use their diverse experiences and backgrounds to eliminate ideas that are not feasible and also develop some consensus on novel ideas that are potentially useful. Explicit instructions to select original ideas or to focus on generating creative ideas appear to increase the tendency to choose more original ideas (Rietzschel, Nijstad, & Stroebe, 2010; Rietzschel et al., 2014).

One problem with the selection process as examined in past studies is that groups or individuals are confronted with sorting through a large number of ideas. Choosing the best ideas from a large set may be rather daunting, and individuals may "satisfice" to make the process somewhat easier. It is possible that if the idea-generation sessions were broken up by periodic evaluation sessions, the evaluation process would be more effective. The evaluation task would be more feasible, and participants would also benefit by a deeper processing of the ideas already generated. This might stimulate additional and more sophisticated ideas in later sessions (for an example, see Putman & Paulus, 2009). Future studies will have to determine if such an approach is indeed helpful for the evaluation process.

Leadership

In real-world teams, the role of leaders appears to be important in fostering creativity. This is not surprising because we have shown that without appropriate task structure and process and high levels of motivation, groups may not perform very well on creative tasks. Thus, leaders may be required to help motivate the group members and to provide guidance throughout the process. The leadership styles that have been found to be most useful for enhancing creativity have focused on these elements (Almquist, Leiman, Rigby, & Roth, 2013b; Hülsheger et al., 2009; Hunter, Bedell-Avers, & Mumford, 2007; Zaccaro, Heinen, & Shuffler, 2009). The one that has gained the most attention is transformational leadership (Bass & Avolio, 1994; Shamir, House, & Arthur, 1993). Transformational leaders provide a shared vision and high expectations, encourage novel approaches, and provide individual consideration and support. However, this type of leadership is not always beneficial. It appears to work best when there is a strong culture of excellence in the organization (Eisenbeiss, van Knippenberg, & Boerner, 2008) and there is some level of educational diversity (Shin & Zhou, 2007). A recent study found that it is important for leaders to be rather explicit in their direction of creative behavior in groups (Carmeli & Paulus, 2015). Although most studies of leadership style and creativity in groups and teams have been done on teams in organizations, a number of studies have demonstrated that transformational leadership style can have similar beneficial effects on performance in short-term electronic brainstorming groups (Kahai, Sosik, & Avolio, 2003; Sosik, Kahai, & Avolio, 1998).

Implications for Organizational Settings

In this chapter, the various recommendations are derived mostly from highly controlled studies in laboratory settings. Most of these studies used college students who generated ideas on assigned tasks; thus, their relevance to teams in organizational settings needs further elaboration. The few controlled studies done in organizations have found results consistent with those from laboratory studies (e.g., Paulus et al., 1995, 2015). An extensive literature on team innovation has also emerged (Hülsheger et al., 2009). Self-reports or other reports of innovation or creativity may not be particularly well related to actual performance (Paulus, Dzindolet, Poletes, & Camacho, 1993; van Dijk et al., 2012), but considerable findings in the team innovation literature are quite consistent theoretically with the findings on creative performance in groups (Paulus, Dzindolet, & Kohn, 2011). Therefore, we are fairly confident that the application of the various insights suggested in this chapter to innovation in work teams or organizations will enhance the overall level of innovation (discovery or generation of high-quality ideas). However, as we have highlighted in this chapter, there are still some major gaps in our understanding of what the best practices should be. Hence, practitioners are encouraged to follow our suggestions and to experiment with them in order to determine what works best in a particular context.

SUMMARY RECOMMENDATIONS

1. Provide a supportive atmosphere that encourages engagement in creativity.
2. Feedback and some degree of social competition may enhance motivation in groups.
3. An initial focus on quantity increases the number of ideas generated and the number of good ideas.
4. The use of electronic or written exchanges of ideas in groups can enhance the number of ideas generated.

5. Alternating group and individual ideation sessions may allow groups to more fully tap their creative potential.
6. Some degree of intellectual, cultural, and background diversity can increase the number and quality of ideas generated by groups.
7. Task structure such as brief breaks, focusing on one category at a time, or prior generation of a range of subtopics can enhance the number of ideas generated.
8. Groups that consist of members who have positive social or group dispositions tend to be more creative.
9. Groups that have leaders who encourage creativity and provide guidance will demonstrate more innovative behavior.
10. Providing groups experiences with effective approaches to group idea exchange enhances their creative performance.

FUTURE DIRECTIONS FOR RESEARCH

Although much has been learned in the past few decades about the brainstorming process, there are still a number of significant gaps in the literature. Although both brainwriting and electronic brainstorming appear to be useful approaches, no study has examined the impact of group size on brainwriting. In addition, although larger groups would increase exposure to a broad range of ideas, they may also slow the idea exchange process and may lead to information overload. More studies are required to determine the overall utility of different ways of brainwriting (e.g., exchanging ideas as they are generated or in periodic review sessions). Although alternating individual and group brainstorming may be beneficial, research needs to determine what the optimal lengths of time should be for these alternating sessions. This may be influenced by the difficulty of the problem in that more difficult problems may benefit more from group sharing than may relatively easy ones. Brief breaks appear to be beneficial for idea generation, but more research is required to determine what should be done during the breaks. Some evidence suggests that exposure to nature may have restorative effects on attention (Berman et al., 2008), but this has not been examined during breaks from brainstorming. Studies appear to support the benefit of group diversity for creativity, but there have been no systematic studies of the effects of expertise or functional diversity on group brainstorming effectiveness. More research is also required on ways to enhance the selection of the best ideas after brainstorming. For example, a diverse group may be better at selecting high-quality ideas because its members have a broader perspective on the potential value of shared ideas. These issues and other gaps in the literature provide a number of promising avenues for future research.

ACKNOWLEDGMENTS

This chapter and related efforts were supported by collaborative grant BCS 0729305 to Paul B. Paulus from the National Science Foundation, which included support from the Deputy Director of National Intelligence for Analysis and collaborative grants from the National Science Foundation (CreativeIT 0855825 and INSPIRE BCS 1247971). Any opinions, findings, and conclusions or recommendations expressed in this material are those of the authors and do not necessarily reflect the views of the National Science Foundation. The authors thank Eyal Nagar for his support during this research project.

References

Almquist, E., Leiman, M., Rigby, D., & Roth, A. (2013a, May 8). Taking the measure of your innovation performance. *Bain Brief*.

Almquist E., Leiman, M., Rigby, D., & Roth, A. (2013b, August 19). Cracking the code of innovation. *Forbes.com*.

Barry, B., & Stewart, G. L. (1997). Composition, process, and performance in self-managed groups: The role of personality. *Journal of Applied Psychology, 82*(1), 62–78.

Barsh, J., Capozzi, M., & Mendonca, L. (2007, October). How companies approach innovation: A McKinsey global survey. *McKinsey Quarterly*, 1–10.

Baruah, J., & Paulus, P. B. (2008). Effects of training on idea generation in groups. *Small Group Research, 39,* 523–541.

Baruah, J., & Paulus, P. (2009). Enhancing group creativity: The search for synergy. In E. A. Mannix, M. A. Neale, & J. A. Goncalo (Eds.), *Creativity in groups: Research on managing groups and teams* (Vol. 12, pp. 29–56). Bingley, UK: Emerald Group.

Baruah, J., & Paulus, P. B. (2011). Category assignment and relatedness in the group ideation process. *Journal of Experimental Social Psychology, 47,* 1070–1077.

Bass, B. M., & Avolio, B. J. (Eds.). (1994). *Improving organizational effectiveness through transformational leadership.* Thousand Oaks, CA: Sage.

Bechtoldt, M. N., De Dreu, C. K. W., Nijstad, B. A., & Choi, H.-S. (2010). Motivated information processing, social tuning, and group creativity. *Journal of Personality and Social Psychology, 99*(4), 622–637.

Bell, S. T., Villado, A. J., Lukasik, M. A., Belau, L., & Briggs, A. (2011). Getting specific about demographic diversity variable and team performance relationships: A meta-analysis. *Journal of Management, 37,* 709–743.

Berman, M. G., Jonides, J., & Kaplan, S. (2008). The cognitive benefits of interacting with nature. *Psychological Science, 19,* 1207–1212.

Bouchard, T. J., Jr., & Hare, M. (1970). Size, performance, and potential in brainstorming groups. *Journal of Applied Psychology, 54,* 51–55.

Brown, V. R., & Paulus, P. B. (2002). Making group brainstorming more effective: Recommendations from an associative memory perspective. *Current Directions in Psychological Science, 11,* 208–212.

Camacho, L. M., & Paulus, P. B. (1995). The role of social anxiousness in group brainstorming. *Journal of Personality and Social Psychology, 68,* 1071–1080.

Carmeli, A., & Azerual, B. (2009). How relational capital and knowledge combination capability enhance the performance of knowledge work units in a high-technology industry. *Strategic Entrepreneurship Journal, 3*(1), 85–103.

Carmeli, A., Gelbard, R., & Reiter-Palmon, R. (2013). Leadership, creative problem-solving capacity, and creative performance: The importance of knowledge sharing. *Human Resource Management, 52*(1), 95–122.

Carmeli, A., & Paulus, P. B. (2015). CEO ideational facilitation leadership and team exploratory behaviors: The mediating role of knowledge sharing. *Journal of Creative Behavior, 49,* 53–75.

Choi, H.-S., & Thompson, L. (2005). Old wine in a new bottle: Impact of membership change on group creativity. *Organizational Behavior and Human Decision Processes, 98,* 121–132.

Collins, C. J., & Smith, K. G. (2006). Knowledge exchange and combination: The role of human resource practices in the performance of high-technology firms. *Academy of Management Journal, 49,* 544–560.

Coskun, H. (2000). *The effects of outgroup comparison, social context, intrinsic motivation, and collective identity in brainstorming groups.* Unpublished doctoral dissertation, The University of Texas at Arlington, Arlington, TX.

Coskun, H., Paulus, P. B., Brown, V., & Sherwood, J. J. (2000). Cognitive stimulation and problem presentation in idea generation groups. *Group Dynamics: Theory, Research, and Practice, 4,* 307–329.

Csikszentmihályi, M. (1997). *Creativity: Flow and the psychology of discovery and invention*. New York, NY: Harper Perennial.

De Dreu, C. K. W., Nijstad, B. A., Bechtoldt, M. N., & Baas, M. (2011). Group creativity and innovation: A motivated information processing perspective. *Psychology of Aesthetics, Creativity, and the Arts*, 5(1), 81–89.

De Rosa, D. M., Smith, C. L., & Hantula, D. A. (2007). The medium matters: Mining the long-promised merit of group interaction in creative idea generation tasks in a meta-analysis of the electronic group brainstorming literature. *Computers in Human Behavior*, 23, 1549–1581.

Dennis, A. R., Valacich, J. S., Connolly, T., & Wynne, B. E. (1996). Process structuring in electronic brainstorming. *Information Systems Research*, 7, 268–277.

Dennis, A. R., & Williams, M. L. (2003). Electronic brainstorming: Theory, research, and future directions. In P. B. Paulus & B. A. Nijstad (Eds.), *Group creativity: Innovation through collaboration* (pp. 160–180). New York, NY: Oxford University Press.

Deuja, A., Kohn, N. W., Paulus, P. B., & Korde, R. (2014). Taking a broad perspective before brainstorming. *Group Dynamics; Theory, Research, and Practice*, 18(3), 222–236.

Diehl, M., & Stroebe, W. (1987). Productivity loss in brainstorming groups: Toward the solution of riddle. *Journal of Personality and Social Psychology*, 53, 497–509.

Dugosh, K. L., & Paulus, P. B. (2005). Cognitive and social comparison processes in brainstorming. *Journal of Experimental Social Psychology*, 41, 313–320.

Eisenbeiss, S., van Knippenberg, D., & Boerner, S. (2008). Transformational leadership and team innovation: Integrating team climate principles. *Journal of Applied Psychology*, 93, 1438–1446.

Feist, G. J. (2010). The function of personality in creativity: The nature and nurture of the creative personality. In J. C. Kaufman & R. J. Sternberg (Eds.), *The Cambridge handbook of creativity* (pp. 113–130). New York, NY: Cambridge University Press.

Goldberg, L. (1992). The development of markers for the Big-Five factor structure. *Psychological Assessment*, 4, 26–42.

Goldenberg, O., Larson, J. R., & Wiley, J. (2013). Goal instructions, response format, and idea generation in groups. *Small Group Research*, 44, 227–256.

Hargadon, A. (2008). Creativity that works. In J. Zhou & C. E. Shalley (Eds.), *Handbook of organizational creativity* (pp. 323–343). New York, NY: Erlbaum.

Heslin, P. A. (2009). Better than brainstorming? Potential boundary conditions to brainwriting for idea generation in organizations. *Journal of Occupational and Organizational Psychology*, 82, 129–145.

Hoff, E. V., Carlson, I. M., & Smith, G. J. W. (2011). Personality. In M. D. Mumford (Ed.), *Handbook of organizational creativity* (pp. 241–270). New York, NY: Elsevier.

Hülsheger, U. R., Anderson, N., & Salgado, J. F. (2009). Team-level predictors of innovation at work: A comprehensive meta-analysis spanning three decades of research. *Journal of Applied Psychology*, 94, 1128–1145.

Hunter, S. T., Bedell-Avers, K., & Mumford, M. D. (2007). The typical leadership study: Assumptions, implications, and potential remedies. *Leadership Quarterly*, 18, 435–446.

Kahai, S. S., Sosik, J. J., & Avolio, B. J. (2003). Effects of leadership style, anonymity, and rewards on creativity-relevant processes and outcomes in an electronic meeting system context. *Leadership Quarterly, 14*(4–5), 499–524.

Keller, R. T. (2001). Cross-functional project groups in research and new product development: Diversity, communication, job stress, and outcomes. *Academy of Management Journal, 44,* 547–555.

Kohn, N. W., Paulus, P. B., & Choi, Y. (2011). Building on the ideas of others: An examination of the idea combination process. *Journal of Experimental Social Psychology, 47,* 554–561.

Kohn, N. W., & Smith, S. M. (2011). Collaborative fixation: Effects of others' ideas on brainstorming. *Applied Cognitive Psychology, 25,* 359–371.

Korde, R. (2014). *Hybrid brainwriting: The efficacy of alternating between individual and group brainstorming and the effect of individual differences.* Unpublished doctoral dissertation, University of Texas at Arlington, Arlington, TX.

Kramer, T. J., Fleming, G. P., & Mannis, S. M. (2001). Improving face-to-face brainstorming through modeling and facilitation. *Small Group Research, 32*(5), 533–557.

Larey, T. S., & Paulus, P. B. (1995). Social comparison and goal setting in brainstorming groups. *Journal of Applied Social Psychology, 26,* 1579–1596.

Larey, T. S., & Paulus, P. B. (1999). Group preference and convergent tendencies in groups: A content analysis of group brainstorming performance. *Creativity Research Journal, 12,* 175–184.

Linsey, J. S., Clauss, E. F., Kurtoglu, T., Murphy, J. T., Wood, K. L., & Markman, A. B. (2011). An experimental study of group idea generation techniques: Understanding the roles of idea representation and viewing methods. *Journal of Mechanical Design, 133*(3), 031008-1-15.

Mannix, E. A., Neale, M., & Goncalo, J. A. (Eds.). (2009). *Research on managing groups and teams: Creativity in groups* (Vol. 12). Bingley, UK: Emerald Group.

Michinov, N. (2012). Is electronic brainstorming or brainwriting the best way to improve creative performance in groups? An overlooked comparison of two idea generation techniques. *Journal of Applied Social Psychology, 42,* 222–243.

Michinov, N., Jamet, E., Métayer, N., & Le Hénaff, B. (2015). The eyes of creativity: Impact of social comparison and individual creativity on performance and attention to others' ideas during electronic brainstorming. *Computers in Human Behavior, 42,* 57–67.

Moreland, R. L., Argote, L., & Krishnan, R. (1996). Social shared cognition at work: Transactive memory and group performance. In J. L. Nye & A. M. Brower (Eds.), *What's social about social cognition? Research on socially shared cognition in small groups* (pp. 57–84). Thousand Oaks, CA: Sage.

Mueller, J. S., Melwani, S., & Goncalo, J. A. (2012). The bias against creativity: Why people desire but reject creative ideas. *Psychological Science, 23*(1), 13–17.

Munkes, J., & Diehl, M. (2003). Matching or competition? Performance comparison processes in an idea generation task. *Group Processes & Intergroup Relations, 6*(3), 305–320.

Nakui, T, Paulus, P. B., & van der Zee, K. I. (2011). The role of attitudes in reactions to diversity in work groups. *Journal of Applied Social Psychology, 41,* 2327–2351.

Nemeth, C. J., & Nemeth-Brown, B. (2003). Better than individuals? The potential benefits of dissent and diversity for group creativity. In P. B. Paulus & B. A. Nijstad (Eds.), *Group creativity: Innovation through collaboration* (pp. 63–84). New York, NY: Oxford University Press.

Nemeth, C. J., & Ormiston, M. (2007). Creative idea generation: Harmony versus stimulation. *European Journal of Social Psychology, 37*, 524–535.

Nijstad, B. A., & Stroebe, W. (2006). How the group affects the mind: A cognitive model of idea generation in groups. *Personality and Social Psychology Review, 10*, 186–213.

Osborn, A. F. (1957). *Applied imagination: Principles and procedures of creative thinking.* New York, NY: Scribner.

Oxley, N. L., Dzindolet, M. T., & Paulus, P. B. (1996). The effects of facilitators on the performance of brainstorming groups. *Journal of Social Behavior and Personality, 11*, 633–646.

Paulus, P. B., & Brown, V. (2003). Enhancing ideational creativity in groups: Lessons from research on brainstorming. In P. B. Paulus & B. A. Nijstad (Eds.), *Group creativity: Innovation through collaboration* (pp. 110–136). New York, NY: Oxford University Press.

Paulus, P. B., & Brown, V. R. (2007). Toward more creative and innovative group idea generation: A cognitive–social–motivational perspective of group brainstorming. *Social and Personality Psychology Compass, 1*, 248–265.

Paulus, P. B., & Coskun, H. (2012). Group creativity. In J. M. Levine (Ed.), *Group processes* (pp. 215–239). Amsterdam, The Netherlands: Elsevier.

Paulus, P. B., & Dzindolet, M. T. (1993). Social influence processes in group brainstorming. *Journal of Personality and Social Psychology, 64*, 575–586.

Paulus, P. B., Dzindolet, M. T., & Kohn, N. (2011). Collaborative creativity—Group creativity and team innovation. In M. D. Mumford (Ed.), *Handbook of organizational creativity* (pp. 327–357). New York, NY: Elsevier.

Paulus, P. B., Dzindolet, M. T., Poletes, G. W., & Camacho, L. M. (1993). Perception of performance in group brainstorming: The illusion of group productivity. *Personality and Social Psychology Bulletin, 19*, 78–89.

Paulus, P. B., Kohn, N. W., & Arditti, L. E. (2011). Effects of quantity and quality instructions of brainstorming. *Journal of Creative Behavior, 45*, 38–46.

Paulus, P. B., Kohn, N. W., Arditti, L., & Korde, R. (2013). Understanding the group size effect in electronic brainstorming. *Small Group Research, 44*(3), 332–352.

Paulus, P. B., Korde, R. M., Dickson, J. J., Carmeli, A., & Cohen-Meitar, R. (2015). Asynchronous brainstorming in an industrial setting: Exploratory studies. *Human Factors, 57*(6), 1076–1094.

Paulus, P. B., Larey, T. S., & Ortega, A. H. (1995). Performance and perception of brainstormers in an organizational setting. *Basic and Applied Social Psychology, 17*, 249–265.

Paulus, P. B., Larey, T. S., Putman, V. L., Leggett, K. L., & Roland, E. J. (1996). Social influence processes in computer brainstorming. *Basic and Applied Social Psychology, 18*, 3–14.

Paulus, P. B., Nakui, T., Putman, V. L., & Brown, V. R. (2006). Effects of task instructions and brief breaks on brainstorming. *Group Dynamics: Theory, Research, and Practice, 10*(3), 206–219.

Paulus, P. B., & van der Zee, K. I. (2015). Creative processes in culturally diverse teams. In S. Otten, K. I. van der Zee, & M. Brewer (Eds.), *Towards inclusive organizations: Determinants of successful diversity management at work* (pp. 108–131). New York, NY: Psychology Press.

Paulus, P. B., & Yang, H. (2000). Idea generation in groups: A basis for creativity in organizations. *Organizational Behavior and Human Decision Processes, 82*, 76–87.

Putman, V. L. (2001). *Effects of additional rules and dominance on brainstorming and decision making.* Unpublished doctoral dissertation, University of Texas at Arlington, Arlington, TX.

Putman, V. L., & Paulus, P. B. (2009). Brainstorming, brainstorming rules, and decision making. *Journal of Creative Behavior, 43*, 23–39.

Rietzschel, E. F., Nijstad, B. A., & Stroebe, W. (2006). Productivity is not enough: A comparison of interactive and nominal brainstorming groups on idea generation and selection. *Journal of Experimental Social Psychology, 42*(2), 244–251.

Rietzschel, E. F., Nijstad, B. A., & Stroebe, W. (2007). Relative accessibility of domain knowledge and creativity: The effects of knowledge activation on the quantity and originality of generated ideas. *Journal of Experimental Social Psychology, 43*, 933–946.

Rietzschel, E. F., Nijstad, B. A., & Stroebe, W. (2010). The selection of creative ideas after individual idea generation: Choosing between creativity and impact. *British Journal of Psychology, 101*(1), 47–68.

Rietzschel, E. F., Nijstad, B. A., & Stroebe, W. (2014). Effects of problem scope and creativity instructions on idea generation and selection. *Creativity Research Journal, 26*(2), 185–191.

Shamir, B., House, R. J., & Arthur, M. B. (1993). The motivational effects of charismatic leadership: A self-concept based theory. *Organization Science, 4*, 577–594.

Shin, S. J., & Zhou, J. (2007). When is educational specialization heterogeneity related to creativity in research and development teams? Transformational leadership as a moderator. *Journal of Applied Psychology, 92*, 1709–1721.

Simonton, D. K. (2004). *Creativity in science: Chance, logic, genius, and zeitgeist.* New York, NY: Cambridge University Press.

Smith, S. (2003). The constraining effects of initial ideas. In P. B. Paulus & B. A. Nijstad (Eds.), *Group creativity: Innovation through collaboration* (pp. 15–31). New York, NY: Oxford University Press.

Snyder, M., & Ickes, W. (1985). Personality and social behavior. In G. Lindzey & E. Aronson (Eds.), *Handbook of social psychology* (3rd ed., Vol. 2, pp. 883–947). New York, NY: Random House.

Sosik, J. J., Kahai, S. S., & Avolio, B. J. (1998). Transformational leadership and dimensions of creativity: Motivating idea generation in computer-mediated groups. *Creativity Research Journal, 11*(2), 111–121.

Stasser, G., Abele, S. P., & Parsons, V. S. (2012). Information flow and influence in collective choice. *Group Processes & Intergroup Relations, 15*, 619–635.

van der Vegt, G. S., Bunderson, J. S., & Oosterhof, A. (2006). Expertness diversity and interpersonal helping in teams: Why those who need the most help end up getting the least. *Academy of Management Journal, 49*(5), 877–893.

van Dijk, H., van Engen, M. L., & van Knippenberg, D. (2012). Defying conventional wisdom: A meta-analytical examination of the differences between demographic and job-related diversity relationships with performance. *Organizational Behavior and Human Decision Processes, 119*(1), 38–53,

van Knippenberg, D., & Schippers, M. C. (2007). Work group diversity. *Annual Review of Psychology, 58*, 515–541.

Watson, W. E., Kumar, K., & Michaelson, L. K. (1993). Cultural diversity's impact on interaction process and performance: Comparing homogeneous and diverse task groups. *Academy of Management Journal, 36*, 590–602.

Zaccaro, S. J., Heinen, B., & Shuffler, M. (2009). Developing adaptive teams: A theory of dynamic team leadership. In E. Salas, G. F. Goodwin, & C. S. Burke (Eds.), *Team effectiveness in complex organizations: Cross-disciplinary perspectives and approaches* (pp. 83–155). New York, NY: Routledge.

Ziegler, R., Diehl, M., & Zijlstra, G. (2000). Idea production in nominal and virtual groups: Does computer-mediated communication improve group brainstorming? *Group Processes & Intergroup Relations, 3*(2), 141–158.

4

Applying the Creative Problem Solving Process to Open Innovation

WAYNE FISHER ∎

Creative problem solving is a proven model for driving innovation when implemented as an organization-wide business process (Basadur, 2001; Basadur & Gelade, 2003). Figure 4.1 is an adaptation of Basadur's Simplex model (Basadur, 2001) that has been used successfully in hundreds of innovation workshops at Procter & Gamble's innovation studio, the GYM. In the Simplex framework, creative problem solving follows three distinct phases: problem formulation, solution finding, and execution. Each phase includes a divergent step (in which all options are explored) followed by a convergent step (in which the most promising options are carried forward to the next step in the process).

Basadur's research shows that organizations with a culture of continuous problem finding, problem solving, and implementation—along with requisite attitudinal, behavioral, and cognitive skills—have the greatest long-term innovation success (Basadur & Gelade, 2003).

By way of example, in the early 2000s, Procter & Gamble (P&G) CEO A. G. Lafley declared that "innovation is everyone's job" and revamped

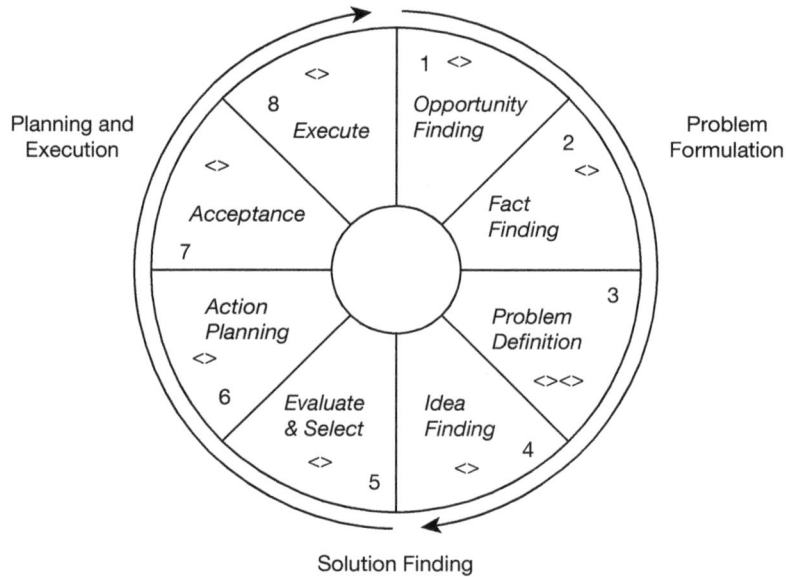

Figure 4.1 The creative problem solving process.

its innovation infrastructure by embracing "design thinking" as a key enabler of the innovation process (Berner & Brady, 2005). Consistent with Lafley's "consumer is boss" business approach, design thinking is a variation of creative problem solving that actively involves the consumer in the innovation process. To support this vision, P&G developed a global network of innovation guides and created "the GYM," an innovation studio where internal teams can come for expert guidance through the creative problem solving process (Soboll, 2011). At the same time, P&G embraced open innovation as a source of technologies to help accelerate the development of new products and services (Mascioni, 2011).

Within this culture, open innovation serves as a useful tool for sharing problems and developing new solution approaches—augmenting the knowledge and experience base of the project team.

However, open innovation, as it is often practiced, violates some very basic principles of creative problem solving. Effective problem definition, in particular, is a known barrier to traditional crowdsourced open innovation approaches (Spradlin, 2012). Creative problem solving is also an inherently iterative practice that benefits from the synthesis of multiple points of view and cycling between problem definition and ideation. These important benefits are lost in traditional point-to-point open innovation involving a lone seeker and isolated solvers.

This chapter explores the most common barriers to effective problem definition and provides practical tools and techniques for facilitating effective problem definition prior to engaging open innovation partners. It also provides a framework for decomposing a complex innovation challenge into focused problem statements that can be solved immediately (using well-known ideation techniques), deployed internally, crowd-sourced externally, and/or addressed over time. Finally, it shares a case study of how open innovation can be enhanced by engaging multiple touchpoints across the supply chain.

RECENT TRENDS IN OPEN INNOVATION PRACTICE

A retrospective analysis of the Product Development and Management Association's Outstanding Corporate Innovator award winners (Kay et al., 2012) shows that open innovation has clearly emerged as a key corporate growth strategy during the past decade. Outstanding Corporate Innovator award winners have developed new practices and processes for open innovation that enhance product development effectiveness and reduce development costs/risk while also managing the complexities of working with external partners.

DSM (a global science-based company active in health, nutrition, and materials) was the Outstanding Corporate Innovator award winner in 2009. DSM made open innovation a fundamental corporate strategy. Its market-driven innovation program follows the classic "porous funnel" model of open innovation (Product Development and Management Association, 2009).

One emerging trend in open innovation is engagement across multiple touchpoints in the supply chain. Kennametal (a global supplier of wear-resistant tooling and industrial materials) was the Outstanding Corporate Innovator award winner in 2010. Kennametal leverages open innovation to deliver value, growth, and productivity solutions to its customers (Product Development and Management Association, 2010). The company's Beyond Blast technology for high-speed milling of titanium addressed an important customer need—a dramatic improvement in both piece rate and tool life. Development of this game-changing technology required breakthroughs across the supply chain—metallurgy, fluid flow and heat transfer modeling, cutting tool design, milling machine design, and "plug-and-play" solutions for manufacturers to upgrade their existing processes.

A MODEL FOR ENGAGING BOTH CUSTOMERS AND SUPPLIERS IN OPEN INNOVATION

The creative problem solving model provides a useful framework for engaging both customers and suppliers in the innovation process. In the traditional open innovation model, it is the responsibility of the seeker to work through the problem formulation steps—opportunity finding, fact finding, and problem definition. The seeker then prepares a written challenge statement for distribution to the solver community. Solvers with relevant subject matter expertise are then invited to offer potential solution approaches (the idea-finding step) to be evaluated by the seeker (the evaluate and select step). The seeker then completes the planning and execution steps—action planning, acceptance, and execute—either with or without ongoing support from the solver.

This point-to-point crowdsourcing approach to open innovation is a proven model for solving narrow, well-defined problems. In practice, however, effective problem definition is often a complex, iterative process. In a typical 2-day problem solving workshop such as those held at P&G's innovation studio, the GYM, participants will generate 100–200 possible problem statements before converging on the top 5 or so to take forward into ideation. The workshop participants have the benefit of understanding the broader context from which the top problem statements were selected, and they have the opportunity to revisit/refine the problem statements as needed during ideation.

Engaging suppliers in the conversation broadens the viewpoint of "what's possible" during problem definition and idea generation. "Voice of the customer" input is a traditional element of fact finding and problem definition. Further engaging customers in idea generation and evaluation improves the probability of eventual acceptance of the products being developed. Setting a lighthouse vision of *"How would we behave if we were one company?"* opens new possibilities for win–win solutions by maximizing value across the supply chain.

Given the complexity of any given supply chain, a comprehensive problem definition exercise including customers and suppliers will result in hundreds of potential problem statements spanning many fields of science. A detailed case study of engaging both customers and suppliers is given at the end of this chapter.

OPEN INNOVATION VERSUS CREATIVE PROBLEM SOLVING/DESIGN THINKING

Open innovation, as practiced in the crowdsourcing model, violates some very basic principles of both traditional creative problem solving and the emerging practice of design thinking. First and foremost, creative problem solving and design thinking are iterative processes. Problem definition in particular is a messy, divergent, iterative process. Design thinking further promotes a "rapid prototyping" mindset in which a potential problem statement cannot be evaluated until it has been used as stimulus for ideation, and ideas cannot be evaluated until they have been prototyped for customer feedback. Here, the objective is to "try on" as many problem statements and solution approaches as possible, while leveraging the collective wisdom of the solving team.

Writing a clear, concise challenge statement for submission to an open innovation solution provider, on the other hand, is inherently a convergent process. The seeker must select the "right" problem and describe it at the "right" level of abstraction to attract the "right" solution that can be implemented with the minimum investment of time, money, and resources.

Historically, open innovation solution providers have focused on identifying potential solvers with breakthrough technologies for their client's new product development. These solution providers found that poor problem definition became a key barrier to traditional open innovation approaches (Spradlin, 2012). Today, solution providers such as InnoCentive and NineSigma have moved beyond these traditional "crowdsourcing" services to become open innovation trainers and guides.

Another success factor for collaborative innovation effectiveness is "space management." One tremendous advantage of having a diverse solving team gathered in a physical space is the use of displayed thinking. Literally, the ability to see "everything at once" enables mental connections not otherwise possible. Again, a posted challenge statement provides only one focused viewpoint into a complex problem, and it does not enable a synthesis of solution approaches across the solving community.

BREAKTHROUGH THINKING BEGINS WITH EFFECTIVE PROBLEM DEFINITION

> *We can't solve problems by using the same kind of thinking we used when we created them.*
>
> —Albert Einstein

In a typical creative problem solving workshop, 50% of the time available is spent on effective problem definition. Thus, for a 2-day workshop, the first day is focused on effective problem definition. Also, the majority of session pre-work is focused on effective problem definition.

This emphasis on effective problem definition accelerates the innovation process in important ways:

- Organizational alignment to the top challenges to be addressed
- Clear distinction between short-term and long-term innovation objectives
- Dissection of complex challenges enabling distribution of work within large teams

- Increased organizational intelligence through cross-training and challenging assumptions

In creative problem solving, the primary objective of problem definition is to provide the stimulus for ideation. Superficial problem statements (e.g., "How to grow sales 20%?") do not stimulate new ways of thinking about old problems. Breakthrough ideas begin with breakthrough problem statements. Min Basadur's (2001) "Why—What's stopping" analysis is one effective technique for generating many divergent problem statements at multiple levels of abstraction. Asking "Why?" ensures that the business context for the problem is understood by all (e.g., "Why is it important that we grow sales at this time?"). It may suggest alternative solution pathways to meet the same business objective. Asking "What's stopping?" often provides the necessary granularity and focus needed to stimulate idea generation ("What's stopping us from growing sales today?"). It is this richness that is often lost in the traditional open innovation process.

OVERCOMING BARRIERS TO EFFECTIVE PROBLEM DEFINITION

Historically, research in creativity has focused on an individual's ability to generate ideas in response to a well-defined problem (Basadur & Gelade, 2003). In practice, however, two clichés ring true: "A problem well defined is a problem half solved" and "Innovation is a team sport." In the absence of effective problem definition, otherwise reliable problem solving tools such as the 40 Inventive Principles from the theory of inventive problem solving (Mann, 2002) will fail to generate breakthrough ideas. The primary objective of creative problem solving facilitation and training at P&G's GYM is to maximizing team, not individual, innovation effectiveness. Much more research is needed in the areas of effective problem definition and team innovation effectiveness, but some preliminary findings are shared here.

One key learning at the GYM is that few individuals or teams can facilitate themselves through problem definition, despite its critical role in the innovation process. Basadur and Gelade (2003) showed that each stage of the creative problem solving process tends to favor a particular cognitive style. A divergent, abstract thinking style is most effective for problem definition. Basadur and Gelade characterize this as a "conceptualizing" style preference. "Conceptualizers" like to think deeply about a problem, explore it from many angles, and generate many alternative solution approaches. On average, "conceptualizing" is the preferred cognitive style for approximately one-fourth of the population. On any given team, this style may be over- or underrepresented.

Basadur's research suggests that unbalanced teams will naturally focus their energy on the steps in the creative problem solving process that match their preferred cognitive style. In contrast to conceptualizers, "implementers" prefer action over contemplation, and they prefer to make choices rather than generate alternatives. Implementers tend to become impatient with conceptualizers during problem definition. Thus, a team that is underrepresented in the "conceptualizing" style and high in "implementing" style will tend to favor quick fixes to any problem that arises. In this situation, deliberate training on the effective problem definition process and skilled facilitation using effective problem definition tools can help implementers overcome their style preference during the problem definition step of creative problem solving.

Even with the correct cognitive style, few individuals are able to do effective problem definition on their own innovation challenge. Art Markman (2012) suggests one potential explanation—the *illusion of explanatory depth*. Simply stated, we do not understand our world as well as we think we do. The more familiar the object (e.g., a Sharpie) or the situation (e.g., our customer's needs), the larger the gap between what we think we know and what we actually know.

To overcome these barriers, the GYM developed an effective problem definition workshop to help seekers prepare challenge statements as part of P&G's Connect & Development program. Even without expert facilitation, Fisher (2012b) found that peer facilitation—along with some basic training in effective problem definition tools and principles—can significantly improve the quality and depth of thinking for a broad range of innovation challenges.

As frequently demonstrated at P&G's GYM, there is tremendous benefit for teams being deliberately facilitated through the creative problem solving process. Experienced facilitators are able to both select the best tools for problem definition and guide the team through their use to a much deeper level of explanatory depth. In facilitated workshops at the GYM, leading teams through an in-depth problem definition exercise will naturally begin to stimulate idea finding—much more effectively than traditional brainstorming approaches. The participants are also well "primed" for idea finding using more advanced ideation tools.

A CASE STUDY: EFFECTIVE PROBLEM DEFINITION

The patent literature provides a rich source of successful problem solving approaches. In fact, the theory of inventive problem solving was developed from an in-depth study of published patents (Mann, 2002).

US Patent 7,413,629 describes a process for producing deep-nested embossed paper products (Fisher, Boatman, Rasch, & Wilke, 2008). Figure 4.2 provides a summary of the key challenges addressed by this invention. The desired outcome is to create a tissue product with improved aesthetics without a significant loss in strength. By exploring the underlying causes of strength loss in traditional embossing processes, we can identify alternative solution approaches to eliminate the historical trade-off between strength and appearance.

Following Basadur's (2001) Simplex method, each new fact uncovered during this exploration becomes a potential new problem to be solved. Ideally, these new problems should be stated in the form of a "How might we?" statement for idea finding.

For example, one reason why strength is lost during the deep nested embossing process is the high strain experienced by the sheet during the embossing process. This high strain is required to overcome the sheet's

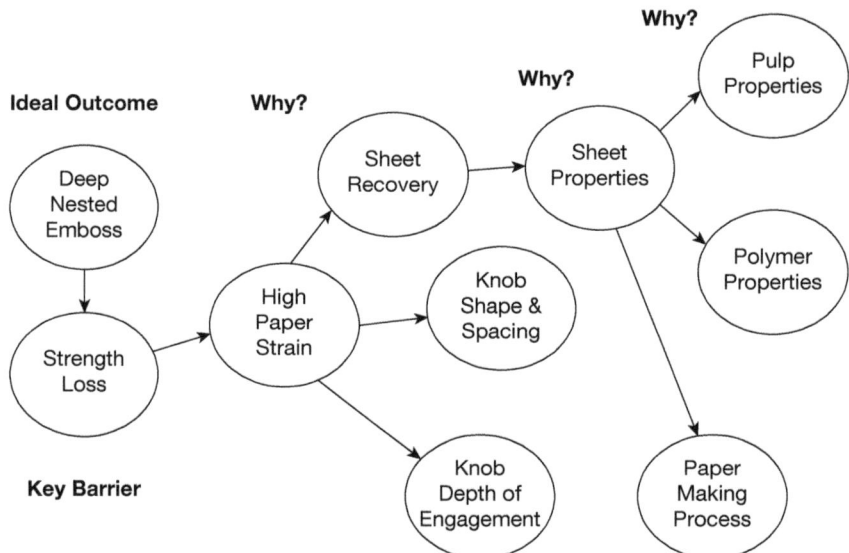

Figure 4.2 Example why–why–why analysis.

tendency to relax and recover its original (flat) shape after embossing. Thus, one potential problem—the one directly addressed by the patent claims—could be stated as "How might we overcome the sheet's natural tendency to recover its flat shape after embossing?"

Further exploration into the causes of strength loss can reveal potential breakthrough problem statements not addressed in the patent:

- How might we reduce the depth of knob engagement needed for embossing?
- How might we change the properties of the pulp fibers?
- How might we change the properties of the wet strength polymers?
- How might we "repair" the damage caused by deep nested embossing?
- How might we "delay" the bonding of the wet strength polymers until after deep nested embossing?

Note that this analysis differs somewhat from the traditional "5 Why" analysis (Ohno, 1988), in which the objective is to find the single "correct" root cause of an observed problem. Rather, this is a divergent process to identify as many problem statements as possible before selecting those to be taken into solution finding. In the previous example, the inventors use steam as a means to modify the properties of the sheet immediately before entering the embossing process. The previous analysis suggests alternative approaches from the fields of structural analysis, paper science, or polymer science that would be excellent candidates for open innovation.

THE INNOVATOR'S TOOLBOX

P&G's global network of innovation guides have compiled a suite of creative problem solving tools for any conceivable innovation challenge (Fisher, 2012b). Fisher (2012a) provides a sample. Tools that are particularly well suited for leveraging open innovation within the creative problem solving framework are discussed here.

The *level of ambition* template (Fisher, 2012a) shown in Figure 4.3 is a useful tool for summarizing all of the potential innovation opportunities available to an organization. It evaluates these opportunities on a plot of

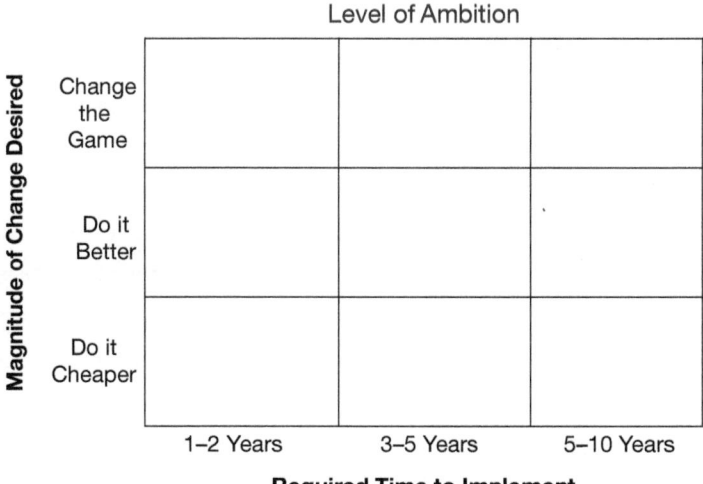

Figure 4.3 Level of ambition template.

"magnitude of change" versus "time to implement." This helps calibrate the team on the magnitude and types of change the organization seeks from innovation and over what time frame. It can also be used to identify which innovation challenges will be addressed internally and which are candidates for open innovation.

Why–why–why or *"Why—What's stopping"* are useful tools to begin the fact-finding process for any type of innovation challenge. As in the embossing example discussed previously, these tools can be used to break down problems into more granular problem statements for idea finding.

Functional analysis/parametric analysis is a particularly useful fact-finding tool for technical problem solving (Mann, 2002). In a training example inspired by the infamous McDonald's hot coffee lawsuit, workshop participants are asked to identify all of the functions and associated parameters involved in brewing, serving, and consuming hot coffee. This analysis typically results in a number of yet unresolved "problems" in the way coffee is prepared and served in fast-food restaurants. Most are unremarkable—for example, "How might we cool the hot coffee after brewing?" and "How might we add the right amount of cream and sugar for each customer's taste?" However, some are more breakthrough and (from McDonald's standpoint) potential open innovation candidates—for example, "How might we brew great-tasting coffee at a safe drinking temperature?" and "How might we design a travel mug that absorbs, stores, and later releases the excess heat from the hot coffee?

The *9 windows analysis* is a fact-finding tool that helps solving teams look beyond the "here and now" and find new ways to look at a problem (Mann, 2002). As illustrated in Figure 4.4, the problem is viewed through two new lenses—time and space. For the McDonald's case, examining the chain of events before and after the actual burn incident uncovers a host of new problem statements—for example, "How might we design a lid that allows cream and sugar to be added without the lid being removed?" Likewise, from zooming in (evaluating system components) and zooming out (looking for resources in the environment)—for example, "How might we use the excess heat from freshly brewed coffee to preheat the water for the next brew cycle?"

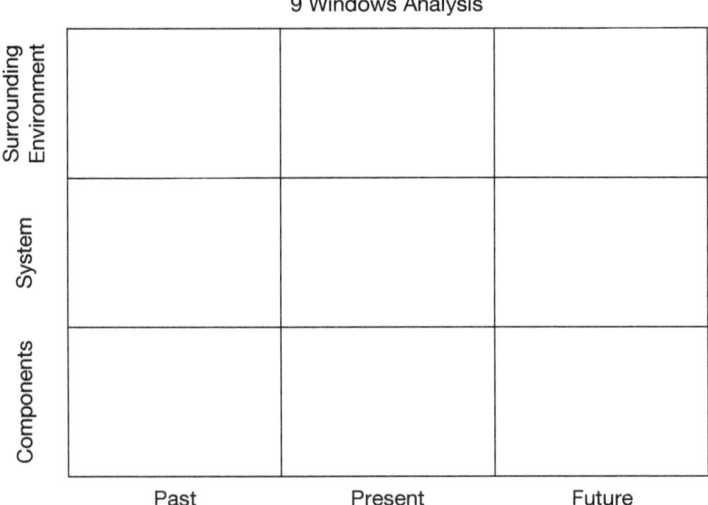

Figure 4.4 9 windows analysis template.

Value curves compare the most important features of a proposed product offering with competitive offerings as a source for new problems to be solved (Lee, 2012).

The *theory of inventive problem solving* or TRIZ provides a rich source of solution approaches for idea finding (Mann, 2002). The 40 Inventive Principles developed by Altshuller (2005) are particularly useful for technical problem solving, but they have been successfully reapplied to many other fields. Although the principles may suggest a breakthrough solution approach, how to apply the principles to a specific process may be another opportunity for open innovation. For example, "Principle 18—Mechanical Vibration" is a well-known way to improve mass transfer in extraction processes, and it may provide a means to brew great-tasting coffee at a safer temperature.

The use of *analogy* is another well-known approach for idea finding (Markman & Wood, 2009). In the context of open innovation, the use of analogy helps the seeker reframe the problem statement in way that the challenge statement can attract solvers from other industries. Spradlin (2012) gives an excellent example of how a chemist in the cement industry provided a breakthrough solution to the Exxon Valdez oil spill cleanup effort (coincidentally using "Principle 18—Mechanical Vibration").

IDENTIFYING CANDIDATES FOR CROWDSOURCED OPEN INNOVATION

The best time to identify opportunities for traditional crowdsourced problem solving is immediately *after* the team has gone through the complete creative problem solving process. Effective problem definition will identify both short- and long-term opportunities for innovation. The problem statements generated in a typical creative problem solving workshop fall in a continuum:

- Problems the team can solve using its own knowledge
- Problems that would benefit from the broader organization's knowledge
- Problems that would benefit from customer and supplier knowledge
- Problems that require knowledge from a broader solver community

In a typical 2-day creative problem solving workshop, participants can address only approximately the five top problems. Understandably, teams have a strong bias toward selecting problems that can be solved in the short term with their own knowledge. Actively engaging representatives from other areas of their own company (another form of open innovation), as well as customers and suppliers, makes additional sources of knowledge feel more "accessible" and encourages more breakthrough problem selection. Nonetheless, there is always a sense that there is "a lot of money left on the table" when the team reflects on the longer term opportunities that it did not pursue.

The level of ambition template (see Figure 4.3) can be used throughout the workshop to chart problem statements and ideas as they are being generated. The best candidates for open innovation tend to be problems with high breakthrough potential but that are left behind because of the perceived time required to solve and implement. Posting these focused problem statements to the broader solver community may uncover breakthrough solutions that dramatically shorten the required time to implement.

A CASE STUDY: ENGAGING CUSTOMERS AND SUPPLIERS IN CREATIVE PROBLEM SOLVING

A large information technology firm sought ways to offset a decline in revenue and profits from its traditional product line. Specifically, the firm was looking to offer a holistic suite of cloud-based services to create the "office of the future" for small to medium-sized companies and organizations—integrating data storage, high-speed Internet, VOIP phones, teleconferencing, and e-mail. A multifunctional design team rigorously followed the creative problem solving as follows:

Opportunity Finding. The team went through a visioning exercise and SWOT (strengths, weaknesses, opportunities, threats) analysis to identify potential innovation focus areas. The SWOT analysis was used to identify the company's internal strengths and weaknesses in delivering cloud-based solutions compared to those of its competitors (including non-cloud-based solutions). It was also used to evaluate external opportunities and threats, including emerging technology trends, potential hardware and software partners, competitive offerings, and potential competitive response to new product introductions.

These opportunities were evaluated for "scale of change" versus "time to implement" using the level of ambition template. After diverging on possible opportunities, it chose to focus on improving the value proposition of one of its emerging product and service offerings for its existing customer base (companies with fewer than 100 employees) and to develop additional offerings to serve larger customers—for-profit corporations, universities, hospital networks, retailers, and local government offices.

Fact Finding. Traditional landscape assessment (market dynamics, competitive analysis, financial analysis, technology trends, etc.) provided useful context for the chosen opportunity and helped identify short-term problems to be solved (to improve the effectiveness of its current product and service delivery). However, much richer customer insights were needed to inform the new product development process.

In preparation for the workshop, each participant conducted an in-depth customer interview and attended a sales call, a product installation, and a service call. In total, the participants experienced approximately 40 unique customer touchpoints prior to the workshop. The most important insights from the customer immersion experiences were immediately captured using an empathy map (Osterwalder & Pigneur, 2010). During the workshop, participants explored the insights from the empathy maps even further using a why–why–why analysis (Fisher, 2012b) to complete the fact-finding step.

Problem Definition. The multifunctional team, along with representatives from the team's key hardware and software suppliers, used the output from the fact-finding step to generate more than 200 very granular problem statements. These "How might we" statements were then sorted and grouped into 20 key themes. The team also used value curves (Lee, 2012) to compare the elements of performance for its current offerings against competitor's offerings. The resulting gap analysis identified critical areas of improvement for potential future offerings—in other words, additional problems to be solved. The "How might we" statements were prioritized using a simple power dot voting technique (five votes per person).

Idea Finding. After engaging in the highly immersive fact-finding and problem definition experiences, the participants were able to quickly generate business-building ideas for new product offerings, improved customer service, cost savings, and process simplification. In total, more than 100 detailed ideas were generated and captured in the form of an idea card summarizing the following:

- The idea name (for easy reference)
- The specific problem it addressed
- A detailed description of the idea
- How it could be implemented

Evaluate and Select. Each of the idea cards was evaluated for ease of implementation and expected impact. The broader industry perspective of the hardware and software suppliers added greatly to the robustness of this evaluation. The idea cards were then sorted and plotted on four foam core boards (labeled "New Products and Services," "Sales," "Install," and "Service") using the ease versus impact grid. The participants then used a simple voting technique to select the ideas to take into action planning.

Action Planning. Working in subteams, the participants synthesized the top ideas in each focus area into a game plan that summarized the following:

- Key next steps to demonstrate the feasibility/business benefit of the top ideas
- Resource requirements
- Success factors for implementation
- Potential barriers to implementation

In this case study, the active engagement of both customers and suppliers greatly enhanced the productivity of the innovation process. The customer immersion experience followed traditional voice of the customer best practices, but the problem definition tools added richness to the customer insights and greatly facilitated idea finding. The hardware and software suppliers provided a broader industry perspective during all steps of the creative problem solving process. The suppliers in return gained a unique perspective into the unmet needs of both their customers and their customers' customers.

SUMMARY AND RECOMMENDATIONS

Open innovation is most effective when implemented as part of an organization-wide innovation program. The creative problem solving process—with specific tools for problem formulation, solution finding, and execution—provides a robust framework for developing innovative products and services. Open innovation should be practiced within these frameworks rather than as a stand-alone program. Actively engaging customers, suppliers, and solution providers (e.g., as participants in an innovation workshop) will greatly expand the scope of "what's possible" at each step in the creative problem solving process. Using experienced facilitators—with deep knowledge of the creative problem solving process and associated tools—will maximize team innovation productivity and effectiveness.

References

Altshuller, G. (2005). *40 Principles: TRIZ keys to innovation*. Worcester, MA: Technical Innovation Center.

Basadur, M. (2001). *The power of innovation* (pp. 299–316). Toronto, Ontario, Canada: Applied Creativity Press.

Basadur, M., & Gelade, G. (2003). Using the creative problem solving profile for diagnosing and solving real-world problems. *Emergence, 5*(3), 22–47.

Berner, R., & Brady, D. (2005, July 31). Get creative. *Bloomberg Business*. Available at http://www.bloomberg.com/bw/stories/2005-07-31/get-creative.

Fisher, W. (2012a). *Creative problem solving toolbox*. Available at https://repositories.lib.utexas.edu/bitstream/handle/2152/18726/oic2012-fisher-suppl-problem-solving-toolbox.pdf.

Fisher, W. (2012b). *The role of effective problem definition in open innovation*. Available at http://repositories.lib.utexas.edu/bitstream/handle/2152/18726/oic2012-fisher-problem-definition.pdf.

Fisher, W. R., Boatman, D. N., Rasch, D. M., & Wilke, I. I. N. J. (2008). *Process for producing deep-nested embossed paper products*. US Patent 7,413,629, issued August 19, 2008.

Kay, S., Boike, D., Fisher, W., Hustad, T., Jankowski, S., Mills, D., . . . Riggs, B. (2012), Lesson learned from outstanding corporate innovators. *PDMA Visions Magazine, 3*, 10–15.

Lee, R. K. (2012). *Value innovation works*. Charleston, SC: CreateSpace.

Mann, D. (2002). *Hands on systemic innovation*. Kortrijk, Belgium: Creax Press.

Markman, A. (2012, May 3). Do you know what you don't know? *Harvard Business Review*. Available at http://blogs.hbr.org/cs/2012/05/discover_what_you_need_to_know.html.

Markman, A., & Wood, K. (2009). *Tools for innovation: The science behind the practical methods that drive new ideas*. New York: Oxford University Press.

Mascioni, M. (2011, February 7). Open innovation at GE. *MIT Technology Review*. Available at http://www.technologyreview.com/news/422626/open-innovation-at-ge.

Ohno, T. (1988). *Toyota production system: Beyond large-scale production*. Portland, OR: Productivity Press.

Osterwalder, A., & Pigneur, Y. (2010). *Business model generation*. Hoboken, NJ: Wiley.
Product Development and Management Association. (2009). *2009 OCI winner: DSM*. Available at http://www.pdma.org/p/cm/ld/fid=325.
Product Development and Management Association. (2010). 2010 OCI winner: Kennametal. Available at http://www.pdma.org/p/cm/ld/fid=325.
Soboll, P. (2011). Innovation: Behind the buzzword. In F. Moss (Ed.), *Innovation: Perspectives for the 21st century* (pp. 241–255). Madrid, Spain: BBVA. Available at http://www.ideo.com/images/uploads/people/Final_article_for_BBVA.pdf.
Spradlin, D. (2012, September). Are you solving the right problem? *Harvard Business Review, 84.*

5

Opportunity Thinking Approach to Open Innovation

Seeing Clearly the Elephant in the Room

PAM HENDERSON, FRANCESCA LORENZINI, AND GREGORY P. POGUE ■

John Geoffrey Saxe passed an Indian folk story into English culture through the poem, *The Blind Men and the Elephant* (1873) (Saxe, n.d.). In this story, blind men attempt to understand an elephant based on their limited perceptual experience. Feeling with their hands, the men understand different parts of the elephant as a water pipe, fan, column, throne, and sword. It is interesting that this same folk story came to Persia in a different form through the poet Rumi (Figure 5.1). In the Persian version, learned men are presented with a similar challenge as in the English version—to grasp the entirety of an elephant with only the light that a single candle can provide. In both versions, the men make the same mistake: Each touches one place on the elephant and thinks that from his limited perspective he understands the whole animal. By doing so, they fail to grasp the full nature of the beast before them, each making incorrect conclusions. Because these items represented their own individual ends,

Figure 5.1 *Elephant in the Dark*, poem by Rumi, in the original Farsi. A 1663 painting by an unknown Indian artist, illustrating a book by Rumi. This book is held in the Walters Museum.

each man could think of various ways the water pipe, the fan, a column, a throne, or a sword could be used—quite innovatively as solutions to their problems. However, an elephant is none of these items, and the insights and potential uses envisioned by the men would ultimately fail when presented with the full animal. An important departure from the English poem is that in the Persian version, the men are seeing, not blind, but their limited perspective hinders their ability to discern the entire animal. Rumi concludes, "If each of us held a candle there, and if we went in together, we could see it." Darkness can be overcome by both light and cooperation.

The Persian story provides an excellent metaphor for the way we handle innovation. It is easy to become enamored with what is immediately understood by our perceptions, interests, points of view, goals, and ends and by doing so ultimately fail to see "the elephant in the room." The full value of an elephant is not limited to the value it can provide as a sword, column, or fan. These perceptions arise from limitations in our investigations and a misunderstanding of "data" when assessing the elephant. Only when the animal is seen in its entirety is it possible to grasp the associated opportunity and employ it to address human problems.

In our innovation-minded culture, the union of an innovative technology and market need is the "elephant in the room." It may be pursued independently as a problem or solution (or in the dark), but to see it accurately, in its entirety, we must develop candles and weld them collaboratively to gain perspective of the whole. The principles of "opportunity thinking" can act as the candles that deepen and broaden an organization's understanding of a market problem—an "opportunity"—and thereby illuminate the solutions that are best suited to return maximum value (Henderson, 2014). When coupled with open innovation, the opportunity thinking approach capitalizes on collaborative strategies across disciplines and organizational structures, treating these interactions as potential sources of knowledge that have the promise of uncovering larger market opportunities.

This chapter explores the integration of open innovation and the opportunity thinking approach as tools to effectively illuminate "the elephant in the room" and thereby structure more effective innovation practices. We begin by discussing the relationship of entrepreneurial alertness to the elephant of opportunity and develop a framework for enhancing an understanding of the holistic elements of the elephant. We then discuss how the innovator can see the potential for innovation not only in his or her own "room" but also in others—hence creating the basis for open innovation. Next, we apply the opportunity thinking principles to the activities of companies, seeing that these principles and open innovation impact commercial outcomes in business cases.

ENTREPRENEURIAL ALERTNESS TO OPPORTUNITIES

The reader might ask, "What exactly does 'opportunity' mean?" For the purpose of this chapter, we use the term *opportunity* in a manner that goes beyond merely starting a business and instead focuses on the introduction of an innovation (i.e., a non-imitative product or service) into the marketplace (Kirzner, 1979; Schumpeter, 1950). To quote Timmons (1994), an opportunity "has the qualities of being attractive, durable, and timely and is anchored in a product or service which creates or adds value for its buyer or end user" (p. 27).

The word "opportunity" actually derives from *ob portu*, which is Latin for "into port." It came about in the late 1600s when Spain and Portugal were racing for the best trade routes. Interestingly, Portugal conducted one of the first recorded open innovation exercises. It incorporated the design of the sail of African sailing ships into its ships—creating the caravel. This faster ship enabled Portugal to sail to new ports, opening up new opportunity. Traditionally, we would associate the term "opportunity" with the market—the port—the people with money and needs. However, this is not what the word was referring to. From the perspective of the entrepreneurs and technology leaders, the opportunity might be the boat, the technology, and the goods on the boat—the vehicle that delivers the opportunity. But the term did not refer to new value that was created. Instead, it referred to the conditions—the wind and tide—that allowed the boat to make it into port because without these, the boat would stall. This illustration is quite helpful because many opportunities stall when the external conditions are not right to create and deliver the right value for the right needs. This can happen, for example, when a company is too early or late to market, when the value it is trying to create costs more than consumers are willing to pay, or when the business model it adopts fails to capitalize on the potential. The history of business is rife with stalled opportunities, not the least of these being how Kodak did not capitalize on the opportunity for digital photography at the right time (Lucas & Goh, 2009). The story of ships, products, conditions, and ports provides a multifaceted picture of opportunity. Opportunity is in fact all three—the need in the market, the value that can be created through invention, and the conditions in the environment that allow the two to come together at the right time.

It is important to distinguish opportunities from "ideas." Ideas are the specific ways we deliver on an opportunity—they are the products and services created to capture the opportunity. They are usually grounded in the experience (pains or innovations) of the entrepreneur or originator, whereas opportunities are grounded in market practice, customer need, and the broader value platforms that can be created through our technologies and business models. Entrepreneurs are noteworthy for their ability to generate ideas and inventions, but it is the successful entrepreneurs that are equally adept at seeing the larger opportunity and developing ideas that capture the maximum potential within that opportunity.

From the literature, it is clear that opportunity is generally recognized and defined through a process that requires thought, time, and action (Hills, 1995; Long & McMullan, 1984). Opportunities appear through probing the continuum between "marketing" and "entrepreneurial activity," which are each subject to situational circumstances (e.g., changing conditions, gaps, and "vacuums" in the market). The essence of an opportunity is the "fit" of a product or innovation to a market opportunity. The "window of opportunity" can close without warning, demanding focused attention and consideration of the importance of time required to define an opportunity. As time passes and ultimately threatens an opportunity, one must employ the appropriate resources and capabilities to identify the opportunity, if it actually exists, and realize it as a market-fitted product or service (Hulbert, Brown, & Adams, 1997; Timmons, 1994). Within the lexicon of innovation literature, the ability of some individuals or institutions to recognize business opportunity and construct meaningful and profitable solutions is commonly referred to as "entrepreneurial alertness" (Baron, 2006; Kirzner, 1997; Shane & Venkataraman, 2000). This "alertness" reflects the observations of some studies, which suggest that successful entrepreneurs think or approach opportunities differently than others. Even if these insights are not a property unique to entrepreneurial activity, the bulk of research on this topic has focused on trying to understand the dominant cognitive behaviors exercised by entrepreneurs when defining and pursuing opportunity (Busenitz, 1999; Gaglio, 2004; Palich & Bagby, 1995; Simon, Houghton, & Aquino, 1999).

Individuals who display the characteristic of entrepreneurial alertness view changing circumstances, events, and time-based opportunities as signals indicating that the rules of the market game (causal chain) have changed and that the framework linking existing resources (means) and their application to produce value (end) has flaws (Kirzner, 1979). An entrepreneur often sees opportunity in the form of a market approach breaking from the existing "means–end" framework—producing a new causal chain in the marketplace providing a new means to realize value ahead or to the exclusion of competition. The manner in which an entrepreneur sees the "gaps" or outdated features within a causal chain and develops a new means–end framework to produce a business opportunity is not addressed by the perception and reasoning literature alone. It is commonly argued that such attributes are driven by heuristics (Gaglio & Katz, 2001).

Heuristics offer advantages in cognitive processing that may explain why entrepreneurs are both alert to opportunity and can form a clear notion of what opportunity "is" and how to formulate a novel means–end framework, whereas others, seeing the same circumstances and experiencing similar events, fail to move with similar market exposure (Gaglio, 2004; Gaglio & Katz, 2001; Tverksy & Kahneman, 1974). Many studies have found positive correlations between entrepreneurial experience and successful opportunity identification (Corbett, 2007; Davidsson & Honig, 2003; Ucbasaran, Westhead, & Wright, 2008, 2009). Experienced entrepreneurs appear to rapidly and effectively use past experience as a pattern in which to see new business opportunity by comparing new experiences against the past, thereby identifying opportunity (Baron, 2006; Baron & Ensley, 2006). This "experience filter" appears to differentiate the experienced from inexperienced entrepreneurs. Of course, it is crucial to understand which elements of experience are most useful for entrepreneurial success.

Experience and the manner in which heuristics are constructed each have their own obvious advantages, but other scholars have noted many challenges that accompany them. Experienced entrepreneurs can use their "filter" in a way that bars all information from being processed or "selects" elements within a pool of information. In addition, they can rely too much on their experience and not seek wider sources of new information from the broader opportunity ecosystem to weigh against or counsel their "experience filter." This type of cognitive entrenchment can lead to what is referred to as a perceptual "rut" (Baron, 1998; Shepherd & DeTienne, 2005; Ward, 2004), which in turn can lead to a "functional fixedness that inhibits the use and combination of information in novel ways due to associations made in a habitual manner" (Gielnik, Krämer, Kappel, & Frese, 2014, p. 350). Behaviors such as the disregard of new information and stereotypical thought patterns may occur in experienced entrepreneurs because they rely heavily on past experience and miss new (and potentially critical) information associated with changing circumstances (Parker, 2006).

Thus, the question arises: How can inexperienced entrepreneurs develop an effective "experience filter" to see patterns and opportunities more effectively? Taking this further, how can experienced entrepreneurs avoid the trap of relying too much on previous experience and run the risk of missing out because of being closed-minded to broader, more novel opportunities? Finally, how will thinking about opportunities more broadly drive larger open innovation endeavors?

HOW TO IMPROVE ENTREPRENEURIAL ALERTNESS

Taken together, studies of entrepreneurial alertness argue that entrepreneurial experience with associated divergent thinking, personality traits, as well as opportunity identification techniques contribute to the success of experienced entrepreneurs (Ardichvili, Cardozo, & Ray, 2003). The findings of this literature beg the question: How can these techniques and methodologies be improved to address and assess new information? Some studies have investigated the behaviors that inexperienced entrepreneurs can employ to moderate their lack of experience. In the study by Gielnik et al. (2014), the active search and thoughtful processing of information were found to reduce the effect of entrepreneurial experience among the subjects studied. Active engagement in new information also enhanced divergent thinking in both the experienced and the inexperienced individuals. Moreover, high levels of information processing appeared to "replace" what was lacking in experience. Whereas previous models focused on entrepreneurial experience and divergent thinking as the basis for entrepreneurial alertness (Baron, 2006; Shane, 2003; Shane & Venkataraman, 2000), this study examined other means of gaining experience (or experience surrogates) and strongly suggested that certain behaviors and actions can compensate for privations in these areas and potentially sharpen both the elements of entrepreneurial experience and divergent thinking in the model (Gielnik et al., 2014). These studies suggest that broad investigation and disciplined information processing have the potential to add tremendous value to the understanding and identification of opportunity. But how can one identify the types of information to draw upon and how to prioritize their investigation and analysis?

SIX SOURCES OF OPPORTUNITY

Various tools have been devised to assist the process of opportunity assessment (Rice, Kelley, Peters, & O'Connor, 2001). For this discussion, we are most concerned with the key subjects that can set the mindset and readiness of an entrepreneur to be alert for opportunity. Review of the literature (Galbraith, DeNoble, Ehrlich, & Kline, 2007; Galbraith, Ehrlich, & DeNoble, 2006; Short, Ketchen, Shook, & Ireland, 2009) indicates six key sources of information that, with appropriate investigation, can act as candles to illuminate opportunities within internal and external ecosystems (Henderson, 2014). These six sources derive from the three parts of opportunity already discussed—the need, value, and conditions. Needs reside in the market but also reside in our environment and in our own organizations. Value can be created through our technologies but also through our brand expression and business models and derive from our organizations. The conditions that allow the needs to be met by value platforms exist across our own organization, our competitors, and the broader environment. As shown in Figure 5.2, the sources include the pathway from technology to market through organizations, environment, business models, and brand expression. Each is described in practitioner vernacular to emphasize the application of theory to practical use. These sources of opportunity provide a more holistic view of the full opportunity potential—the larger elephant.

Market

The market is often cited as a primary definer of opportunity (Galbraith et al., 2006, 2007). Although market fundamentals such as total available size, growth characteristics, fragmentation, and other characteristics are crucial, the addressability of the market and the meaningful expression of customers' actual needs are key (Moore & McKenna, 1999). Furthermore, it is necessary to consider the manner in which a solution can reach or be distributed to customers. Today, retail instruments are only one of many options available; the increased usage of computers, phones, and (soon) wearable technologies is deepening potential product reach to more consumers. These technologies offer disruptive strategies for current means–end frameworks. Thus, a corollary to Timmons' (1994) definition of market is that opportunity is often found at the interface of market need and the ability to deliver to the market.

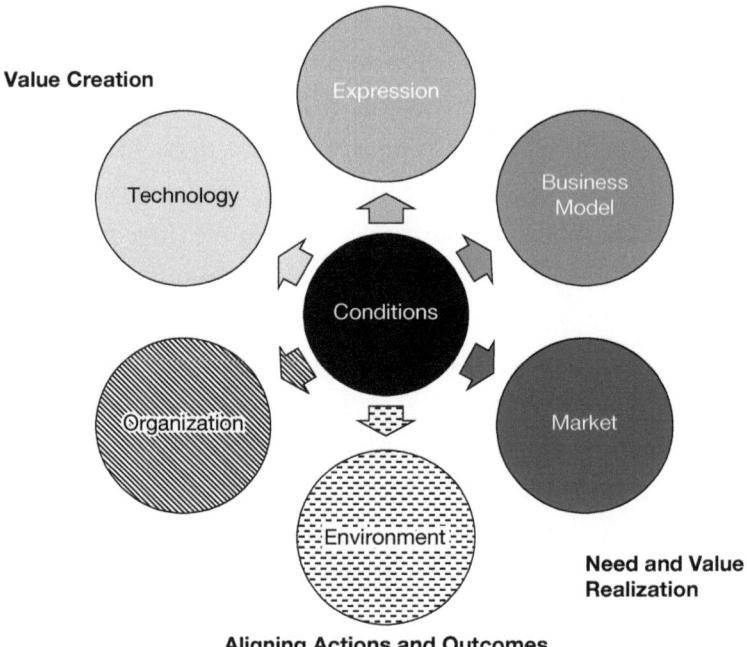

Figure 5.2 The Six Sources of Opportunity and relationship to conditions that unite each. The start of the process is often technology, from which we create value and then end with market—the source of the need and place where we realize value.

Technology

Technology is central to the value offering: The differentiated benefits of the tech are key to customer acquisition, value chain integration, and value realization (Galbraith et al., 2006, 2007; Moore & McKenna, 1999; Rice et al., 2001). Technology can be evolutionary, improving a product's usability, cost, or efficiency, but it can also be revolutionary—creating new products that awaken dormant needs in the market. However, technology often flows from the entrepreneur as a result of direct experience, and because of this attachment, it requires validation outside the entrepreneur (in the market). Otherwise, technology is reduced to a mere solution in search of a problem. Based on this rubric, technology goes beyond what may be protected by intellectual property and includes a company's or entrepreneur's know-how, the processes required to turn a technology into products, and the services that deliver value to customers (Weinstein, 1994).

Organization

Opportunity is rarely recognized and pursued solely in the mind of the entrepreneur. Usually, opportunity is pursued within the context of an organization. Entrepreneurial alertness to opportunity occurs in startup ventures, established companies, government structure, and not-for-profit organizations, all with their own legal guidelines or barriers, organizational structures, and market objectives (Knight & Tamer Cavusgil, 1996; Pinchot & Pelham, 1999). The organization's communication processes, reporting structures, empowerment or reward systems, and employee complement provide the context surrounding the realization of opportunity value. Ultimately, organizations are the source of revenue for a technology. For this reason, they must be considered as sources of opportunity.

Expression

Products and organizations are wrapped into brands comprising messages, designs, market positions, and communication strategies that are meant to garner emotional attachment and brand loyalty in customers (Gruber, 2004; Merrilees, 2007). Key factors of brand management focus on building an identity in the marketplace and demonstrating credibility among potential customers. The first and chief action in marketing and branding strategy is convincing customers of value and encouraging their future likelihood of endorsing the product or service via word of mouth, social media, and by additional purchases. The manner in which organizations communicate the value of a technological innovation depends on their fit into the causal chain and who recognizes the means–end framework advantage of the innovation. These two crucial factors depend on the environment surrounding the market and other contextual factors that influence how customers receive and react to innovative products.

Environment

The environment in which an innovation is produced and commercialized is made up of many things, including the oversight of government—its laws, regulations, and economic policy; culture—trends, values, and media tools; and human influencers, opinion leaders, activists, celebrities, and bloggers. All of these entities form the context in which an innovative product is marketed (Gruber, 2004; Moore, 1991). Entrepreneurs often view their environment as being situated within their local culture or nation of origin due to cognitive entrenchment, which can often lead to perceptual "ruts." However, the "fit" of an innovation with an environment must be viewed broadly, based on its suitability to causal chain interdiction and the potential to generate a new revenue opportunity. International trends, human geographical movements, and multinational activities must each be weighed to identify the most receptive environment for the introduction of a new product.

Business Model

Ultimately, insights from market appetite (and structure), differentiated benefits of technology, organizational values, brand expression, and environmental context must be brought together into an "actionable business model." There is a continuum of literature using this term, with meanings ranging from a business action/role model or plan to an innovation value and a market fit definition (Osterwalder, Pigneur, & Tucci, 2005). Although these may appear disparate, they are each centered around producing an expression of business logic, which relates innovation to customer value, to operational strategy, with financial consequences. In this conception, individuals or groups that benefit from the product, service, or business model are at the core of opportunity (Chesbrough, 2010). The needs, behaviors, and traits of individual consumers are the focus of consumer-focused companies. In contrast, the broader strategies of entire organizations, companies, are the focus of business-to-business entities. Not-for-profits and government-focused organizations consider the community, government policy goals, and often transnational needs as their core. Thus, company goals must link with operational strategies to deliver a product to a potentially receptive market and realize financial value.

Armed with the Six Sources, entrepreneurs have a schema that can direct and aid active information acquisition efforts. By forcing themselves out of comfortable or routine thinking patterns and other ruts and into an alert mindset that can shape opportunity through a more objective framework. This holistic view of opportunity, dictated by the application of the Six Sources, assists entrepreneurs in their escape from limited observations and hasty conclusions about swords, columns, and fans to see the opportunity in its entirety—clarifying the elephant in the room.

DEVELOPING A CIRCUMSPECT VIEW OF THE POTENTIAL FOR OPEN INNOVATION

At this point, it should be clear that the Six Sources do not stand alone and that the deepest insights can be gained by their synergies. Such is the case with open innovation. Open innovation reflects a combination of overall environmental change but also peculiar attitudes and approaches of organizations with regard to innovation practice.

External Practice

Throughout the first half of the 20th century, innovation happened largely within the confines of single institutions. Major organizations orchestrated large innovation projects, for which they would acquire all the financial, technological, and human resources necessary to develop a new technology, integrate the technology into products, and, finally, to commercialize the products. A few industry leaders with access to great resources, such as Bell Labs, DuPont, IBM, and Merck, used internal research and development as the source from which they curated their expansion into new markets (Chesbrough, 2003a, 2006). Over time, changes in government policy (Loise & Stevens, 2010), such as the Bayh–Dole Act, the advent of Internet connectivity, and more economical ways of reaching the market have emerged that do not require companies to acquire full vertical capabilities and at the same time sustain costly in-house research facilities (Chesbrough, 2003b, 2006). Innovations do not become products without significant investment. Growth in capital to finance early stage ventures from business angels, venture capital organizations, and corporate-sponsored research allows early stage entrepreneurs to take creative ideas (which are the raw material of innovation) and turn them into products, allowing for financial reward to be shared by inventor and investor alike (Dudley & Hubbard, 2004).

Thus, the innovation process has been democratized during approximately the past 40 years—with the rise of the applied conceptual expertise within the research university, the availability of risk financing, and the advent of the Internet era. What this has amounted to is that organizations that were either at a disadvantage or an advantage regarding their isolated position within the former "closed" environment are now subject to a new reality, with new roles. Smaller, more nimble organizations have become the engines of innovation. Meanwhile, large companies have restructured to become more agile by looking outside their four walls to other entities, using a more "open" approach: integrating parts from disparate partners into a single marketable product (Chesbrough & Appleyard, 2007; Dodgson, Gann, & Salter, 2006 Quinn, 2000).

These changes have been accelerated by the Internet making the world much smaller, allowing collaboration, co-invention, and co-marketing to become a more mainstream practice (Easley & Kleinberg, 2010). Today, there are billions of computers connected to the Internet (Ryan, 2010). The speed and volume of information transmitted are increasing at an accelerating rate. This new combination of speed and volume has produced new platforms for commerce and social exchange by decreasing the cost of commercial transactions and social interactions. During a short time span, companies have gone from not selling anything over the Internet to creating large and new efficient markets to facilitate exchanges. In a minority of cases, this consisted of extant companies transitioning to new digital platforms; however, a sizeable majority of the innovation consisted of the formation of new companies and capabilities. The Internet, computing, and new capabilities to send and receive sizable amounts of data have as their foundation the formation and value of networks, horizontal structures, and the vast distribution of information. The Internet altered the ways production systems were conceived by allowing deeper integration and more adaptable operations. In addition, instead of ordered, linear methods of conducting research and development, the Internet has facilitated innovation on a concurrent basis. The capacity to expedite the exchange of intelligence and decentralize decision-making within organizations has given rise to more collaborative work environments that do not have geographical constraints. Consequently, former models of centralized internal innovation have yielded to startups and other open innovation structures that better utilize the environment of more ample information. As a result, the new innovation landscape can be better characterized as a knowledge-intensive, rather than resource-intensive, environment (Evans & Annunziata, 2012).

In the open innovation paradigm, new innovation projects are not controlled by a single company or group but, rather, are cultivated within an ecosystem of "role" players: The team that originates an innovative idea is often not the same as the one that funds its development, finalizes its development, or commercializes the technology (West & Lakhani, 2008). The effort of two or more organizations addressing the same market problem in a collaborative manner saves resources and speeds development. In addition, as companies grow and evolve, they may find that improvements to existing products, as well as new ideas, benefit from an entire ecosystem to support their commercialization (Chesbrough, 2006). Thus, in the 21st century, innovation has evolved from a focus on single organizations and tightly controlled projects to a more dynamic process driven increasingly by smaller and newer companies and individual research laboratories (Chesbrough, 2006). Time has shown that the power to innovate has been "spread out" and is now shared by small firms, research institutions, and well-established corporations.

Internal Practice

Although the external ecosystem between organizations provides a valuable source of insights and possibilities for the collaborative development of innovations, the internal ecosystem within an organization can be crucial. If only a single division of an organization is assigned the responsibility of innovating, it can fail because it is not sufficiently aware of the full scope of the actual opportunity within a market. Exchanges of ideas, information, and insight within an organization can help the entity develop its notion of opportunity collaboratively with its employees and offer more efficiency to the innovation process. The opportunity thinking approach encourages leaders to view every division as a contributor to and source of opportunity.

Although it is agreed upon that diversity in hiring practices offers a richer set of experiences and relationships within an organization (Chen & Huang, 2009), it is important to note that organizations can facilitate horizontal relationships inside an organization, opening the firm's environment to new sources of opportunity. Individuals within the divisions and ranks of organizations bring different approaches to their tasks based on their own personal history of participation in the marketplace. For example, technologists will see the cutting edge where innovative advances can drive opportunity. Manufacturing will see opportunity in cost reductions. Business leaders will see bounty in attractive acquisitions.

The same methods that work to gain customer insights can be used to pull insights from an organization's staff. The individuals who constitute every function within companies can practice their own form of openness to generate new ideas and to expand their own opportunities within the organization. However, mechanisms to "listen" and integrate diverse points of view are critical. New insights can be gained through open practices that meld perspectives from different departments and reduce vertical space between management and implementation (Clegg, Unsworth, Epitropaki, & Parker, 2002; Ramus, 2002). Horizontal communication strategies offer the chance to build open innovation ecosystems within organizations, creating a more collaborative and vibrant culture while simultaneously accelerating opportunity realization.

Even with access to broader collaborations, more information, and new, more cost-effective product development functions, the question of "What to develop?" still looms. The subject turns back to opportunity and the alertness of entrepreneurs within and outside of organizations to perceive, measure, and convincingly articulate opportunity into a process capable of delivering an innovative product to the marketplace. The Six Sources can provide the disciplined structure for intensive investigation and information acquisition that give organizations advantage when identifying opportunity. The value of the Six Sources becomes even more apparent when they are applied in an open innovation environment. The holistic perspective gained by using the Six Sources allows entrepreneurs within and outside the organization to see the broader opportunity. In doing so, they can explore what might be called "open opportunity"—a broad view of potential open innovation relationships in the context of opportunity—before a specific relationship is pursued. Open opportunity provides the landscape of possible connections and gives way to larger, more impactful relationships.

APPLYING THE SIX SOURCES TO EXPLORE OPEN OPPORTUNITY AND OPEN INNOVATION

A rich example of exploring open opportunity to produce richer open innovation can be found in an unlikely place. Eastman Chemical Company is a very old organization: It was founded in 1920 and for most of its existence was a part of its parent, Eastman Kodak. The specialization of Eastman Chemical Company was the manufacturing of chemicals, fibers, and plastics. Of its products, cellulose acetate and cellulose triacetate were dominant. Although these products had potential uses in fiber-forming materials and explosives, the most interesting application for Eastman Kodak was the chemicals' use as the base material for photographic emulsions (Ram, 1990). Its development provided a safe replacement of the flammable nitrate film as a film base. In addition, the reaction of materials with camphor and nitrocellulose yields celluloid, which is the basis of the motion picture industry. As technology evolved, the film and motion picture industry began to change, and the demand for these solid products waned in the face of electronic substitutes (Lucas & Goh, 2009). When Eastman was spun off from Kodak in 1994, the company sought new applications for its principal ester and acetate products.

The process Eastman Chemical Company employed in the face of this challenge illustrates the uses of the Six Sources. Eastman quickly took advantage of its technology resources and offered advanced chemistry to transform cellulose precursors into highly derivative and broadly useful chemicals (Pam Henderson, personal communication). Furthermore, the organization possessed scalable manufacturing capabilities to produce these products at a cost and supply level that could support broad industry sectors. The evolving nature of the plastics and film markets provided clarity that the days of their "bread-and-butter" products were on the decline. To hold market value as a stand-alone company, new markets needed to be found in which the applications of Eastman's technologies could serve as the basis of new products (Andal-Ancion, Cartwright, & Yip, 2003; Edgar et al., 2001; Puls, Wilson, Hölter, 2011; Roy, Semsarilar, Guthrie, & Perrier, 2009).

But how should one understand the best "new" applications for an almost century-old product and process? A traditional approach might be to search on a case-by-case basis how large manufacturers of consumer and industrial goods could use these materials. Each collaboration developed would be a means to identify new applications of cellulose derivatives that addressed market needs—but this is time- and cost-intensive. As applications were being sought, Eastman realized through its involvement in the market that the manner in which technologies were being introduced into the market had changed. Technological advances in materials were occurring at an accelerating pace—faster than the company's marketing division could keep up with. This resulted in a lack of understanding for end users regarding the technical capabilities of Eastman's products as they were presented in the company's marketing or expression documents. Thus, the environmental evaluation required understanding the arising value chain—how to discover new uses of Eastman's materials and apply its materials to other emerging technologies into a marketplace largely devoid of scientific knowledge. The company's expression had to change to communicate the value of science in a way that was understandable for those in charge of integrating the company's materials into final products.

Rather than focus on the market in an arduous application development process, Eastman turned its attention to the environment. Obviously, the environment around Eastman's products had changed, which opened up new ways of communicating. Eastman chose an open, horizontal business model so as to more rapidly team up with a broad set of collaborators, scientists, engineers, designers, universities, and consumer goods companies (Andal-Ancion, Cartwright, & Yip, 2003). The challenge Eastman faced, however, was a lack of familiarity in the environment with the properties of the materials and a lack of chemistry expertise to quickly grasp the properties and then develop applications. To address this gap, Eastman developed a new sampling technique that took the cellulose esters from its industrial pellet form and created physical demonstrations of the properties. In doing so, Eastman created a new business model of "selling" its materials to influencers rather than customers—a new approach to Eastman's expression. For example, the new business model involved engaging designers from outside the company to experiment with using the materials in new products. This open innovation activity was in fact different from many in that it utilized the broader opportunity ecosystem of designers to change understanding of the materials and then use their circles of influence to accelerate commercialization. Engagement with designers extended beyond novel sampling techniques to design contests in universities and involvement with the Industrial Design

Society of America. Focusing on open opportunity and benefiting from open innovation revolutionized the organization. Eastman collapsed the distance in the value chain as innovators to the consumers and improved the market uptake of its products. The result was the integration of applications of cellulose ester products from films and cigarette filters to highly derivative applications of novel coatings, printer inks, controlled dosage forms of pharmaceutical products, and even LCD screens (Eastman Chemical annual report, 2007).

The timeliness of the identification of new opportunity and use of open innovation methodologies cannot be overemphasized. The Eastman Chemical Company was sold by Kodak to generate cash, and its organizational management used this newfound environment to build an innovative business model exploiting new market opportunities for derivatives of its core technology creating new brand expressions, thought leadership, and economic success. The Eastman Innovation Laboratory, which was the entity within Eastman Chemical Company that engaged with designers as a result of this strategy, won prestigious industry awards and became the gold standard for marketing due to its new approach to engaging with designers. Moreover, this platform has become a high-growth sector with operating earnings increasing by more than 50% in 2013 compared with 2010, with an operating margin between 10% and 15%. Whereas its parent company declared bankruptcy in 2012, Eastman Chemical has grown to be a *Fortune* 500 company with 14,000 employees at more than 40 manufacturing sites globally yielding products selling in excess of $9.4 billion annually. Applying the Six Sources of Opportunity gave it a market advantage through understanding, pursuing, and exploiting opportunity.

PRACTICING THE OPPORTUNITY THINKING APPROACH IN AN OPEN INNOVATION CONTEXT

From the previous discussion of the Six Sources of Opportunity, the value offered by the opportunity thinking approach when applied in an open innovation framework becomes obvious. How do we realize the value offered? There are five phases to integrating opportunity thinking principles and innovation to realize value (Henderson, 2014):

Phase 1: Ecosystem insight/opportunity landscaping
Phase 2: Selecting which opportunity
Phase 3: Stretching the opportunity
Phase 4: Co-creating the big idea
Phase 5: Market results

Phase 1: Ecosystem Insight/Opportunity Landscaping

As previously discussed, opportunities are sensed by entrepreneurs who are alert to outdated causal chains and the presence of potentially new means–use frameworks extant in the continuum between market and technology function. These are often situational, in which "gaps" in market or time surrounding regulatory and other market features are shrinking. These insights must be identified and landscaped based on their potential to be productively exploited by those with appropriately fitted recognition and resources (Hulbert et al., 1997; Timmons, 1994). This ultimately involves the active search and thoughtful processing of larger amount and variety of market and product information to define the borders and value of opportunity.

The question of "where opportunities lie" requires investigation involving a review of customer practice, geographic strategies and advantage, potential pitfalls, and market dynamics. Which opportunities are most attractive? Which fit best with one's company's goals? As PepsiCo looked for new opportunities to extend its beverage products in an effort to forge a sizable presence in a new market, the company's vision extended from the United States, European Union, and South America to Asia. India was seen as an attractive beachhead to this Asian market. As PepsiCo explored the opportunity, the company saw both a large potential market, with more than 1 billion possible consumers, and numerous challenges, including establishing, extending, and maintaining the product "cold chain,[1]" the business model for delivering beverages into established businesses, and also reaching the consumers frequenting street markets (Pam Henderson, personal communication). PepsiCo successfully launched three new mango drinks as a result of this landscaping work. The company saw broader opportunities to leverage trends in provenance and authenticity and chose to create mango beverages unique to specific mangos within regions. In doing so, PepsiCo defied conventional thinking that leans toward creating highly homogenized fruit beverages to beverages that celebrate subtle taste differences. The company also was open to the possibility of scarcity of supply as it increases perceptions of authenticity. This landscaping process is key to understanding the challenges associated with an opportunity: Together with this market insight, it can aid the formation of a clear idea of who to reach out to for assistance internally and externally to develop the market opportunity.

Phase 2: Selecting Which Opportunity

The process of "selecting" which opportunity to pursue is an action that requires the key properties of entrepreneurial alertness to opportunity: the "experience filter," divergent thinking, and unique approaches toward risk that allow ideas to transition into opportunities. As noted previously, these abilities are often grown heuristically (Gaglio & Katz, 2001) through practice by the entrepreneur. Through this "heuristic" filter of experience, many potential markets that are revealed by the landscaping process start to be ascertained as unequal, thereby making selection possible. Deploying the Six Sources requires energy, resources, and the opportunity costs of rejecting or deselecting potentially promising opportunities to pursue. For these reasons, selection must be made wisely. The process of selection involves fit analysis, which takes into account technical and market challenges, demand assessment, and partnership exploration. Opportunities present themselves where market challenges are able to be addressed by internal capabilities or those found through open innovation partnerships. In addition, market demand must justify the time and resources that are dedicated toward realizing the opportunity.

Phase 3: Stretching the Opportunity

Opportunities taken at face value may be deceptive. On the surface, they may seem relatively insignificant. Opportunities must be stretched to fit the breadth defined by market need and consumer habit (as defined by market segmentation and targeting practice). The point at which The North Face gained access to a proprietary self-wicking fabric technology is a good representation of this phenomenon. The self-wicking technology was one among many options for consumers. The company understood that the features and benefits most relevant to consumer segments and most differentiated from competition were key to understanding the market opportunity and, on this basis, developing their commercialization strategy (Pam Henderson, personal communication). Input from ultra trail runner Lizzie Hawker was critical to this process (Duxbury, 2012; The North Face, *Elizabeth Hawker*, n.d.). Lizzie relayed her common experience of getting cold during rests as her sweat dried too slowly. For The North Face, this experience indicated that there were unmet needs among long-distance runners and that this demographic could be targeted as the company's first market segment. To approach this specific market, The North Face created a FlashDry fabric that the company used in a variety of layering pieces (The North Face, *FlashDry*, n.d.). With this new understanding, the company made the choice to investigate the broader opportunities surrounding the benefits created by a new technology provided by a small technology company. They co-explored the opportunities with that technology company. Only after the two had formed the opportunity landscape and stretched it did they form a joint development agreement for developing specific products. By exploring and stretching the larger opportunities together, the companies formed a more circumspect and expansive partnership.

Phase 4: Co-Creating the Big Idea

Originally viewed as a line extension strategy, the fabric technology was focused toward relatively small market segments that valued moisture management and temperature regulation during outdoor activity without adding significant weight or bulk. The collaboration between The North Face and the technology company allowed for the larger co-creation of a broader value statement. The FlashDry technology was initially targeted at highly competitive long-distance runners who are subjected to changes in weather conditions and require cover in the form of a jacket or other outerwear. The lightweight composition and design of the fabric that did not impede motion allowed these athletes to focus on the trail and competition and not on fighting off chill, wind, rain, or heat. Co-creation involved moving beyond the features and benefits of the technology to specific integrations into their first products, which was the starting point for determining price points, establishing margins and returns, designing their financial strategy, and ultimately served as the source of inertia to drive their initial product to a point of market acceptance.

Phase 5: Market Results

Market results are critical. The company's choice to expand its FlashDry technology into broader applications, from long-distance running to snow sports, climbing and hiking apparel, base layers, and accessory product verticals eventually led to more than 100 products featuring FlashDry technology on the market. FlashDry is a fundamental part of the $3 billion revenue trek that The North Face was working to achieve by the end of 2015 (Duxbury, 2012). As the vice president of marketing at The North Face stated, the use of the opportunity thinking approach took "a concept we thought might have legs and developed it into a big opportunity and then that big idea has taken off with consumers" (Pam Henderson, personal communication).

CONCLUSIONS

The opportunity thinking approach provides a disciplined process to assist entrepreneurs to avoid the pitfalls of experience and cognitive ruts by encouraging the intensive investigation of new information in six key areas. The Six Sources provide a conceptual tool to organize and direct information-gathering exercises. These sources act as candles to deepen an organization's understanding of market opportunity, strengthen the fit of technological offering to market need, and design the best business strategy to unite product with market for mutual benefit. Without this holistic and disciplined approach, entrepreneurs will lean on experience and often miss important information or jump to conclusions based on limited data sets, leading to inaccurate or limited interpretations of the elephant in the room. By applying all Six Sources, the entire elephant can be ascertained, and effective commercial realization of innovation is possible.

The exploration of opportunity also takes the entrepreneur on a journey of exploring potential for innovation both internally and externally. By recognizing the broader framework of opportunity, innovators will be driven to think in terms of the potential for open innovation. The Six Sources further enrich this exploration by driving business models focused on establishing the best collaboration conversations, developing legal linkages, and launching joint activities between sources both outside an organization and within an organization. Open innovation plays an important role in facilitating collaboration and action in this context. In the open innovation framework, an opportunity is the point of focus, which requires the synthesis of perspectives from employees within different divisions of a company along with necessary collaborators. A reduction in hierarchical communications can encourage inter- and intraorganizational collaboration.

Whereas open innovation offers a strategy for collaborative product development, the opportunity thinking approach argues that the participants take a step back, define the opportunity first, and then define roles for interaction to ensure that a complete solution can not only be crafted but also delivered effectively to the market. The approach of considering open opportunity before moving toward specific open innovation engagements can help bolster the understanding of an organization's capabilities and refine the ways they are articulated into the lexicon of other players in the ecosystem that houses it. The examples of Eastman and The North Face offer excellent cases of how open opportunity and open innovation approaches that deploy the Six Sources can illuminate the "elephant" of opportunity in the room. Furthermore, they show how proper understanding of the information gleaned from an environment (or market pain) can lead to more effective product position, marketing, sales, and financial performance. The synergy between open innovation and the opportunity thinking approach provides strong tools for organizations to apply to their activities that can aid their efforts to achieve maximum value and competitiveness in the long term.

ACKNOWLEDGMENTS

We thank Rosemary French and Keela Thomson for their critical review of the open innovation literature and initial summations of its tenants and implications. Dr. Darius Mahdjoubi is acknowledged for introduction of the Rumi poem to the authors and associated discussions. The authors also thank Dr. Art Markman for his critical review of the manuscript and comments.

Note
1. In the industry, this is understood as maintaining a product below room temperature to preserve product quality and expiratory dates.

References

Andal-Ancion, A., Cartwright, P. A., & Yip, G. S. (2003). The digital transformation of traditional businesses. *MIT Sloan Management Review, 44*(4), 34–41.

Ardichvili, A., Cardozo, R., & Ray, S. (2003). A theory of entrepreneurial opportunity identification and development. *Journal of Business Venturing, 18*(1), 105–123.

Baron, R. A. (1998). Cognitive mechanisms in entrepreneurship: Why and when entrepreneurs think differently than other people. *Journal of Business Venturing, 13*, 275-294.

Baron, R. A. (2006). Opportunity recognition as pattern recognition: How entrepreneurs "connect the dots" to identify new business opportunities. *Academy of Management Perspectives, 20*(1), 104-119.

Baron, R. A., & Ensley, M. D. (2006). Opportunity recognition as the detection of meaningful patterns: Evidence from comparisons of novice and experienced entrepreneurs. *Management Science, 52*, 1331-1344.

Busenitz, L. W. (1999). Entrepreneurial risk and strategic decision-making: It's a matter of perspective. *Journal of Applied Behavioral Science, 35*(3), 325-340.

Chen, C. J., & Huang, J. W. (2009). Strategic human resource practices and innovation performance: The mediating role of knowledge management capacity. *Journal of Business Research, 62*, 104-114.

Chesbrough, H. (2006). *Open business models*. Boston, MA: Harvard Business School Publications.

Chesbrough, H. (2010). Business Model Innovation: Opportunities and Barriers. *Long Range Planning, 43*, 354-363.

Chesbrough, H. W. (2003a). A better way to innovate. *Harvard Business Review, 81*(7), 12-3.

Chesbrough, H. W. (2003b). *Open innovation: The new imperative for creating and profiting from technology*. Boston, MA: Harvard Business School Publications.

Chesbrough, H. W., & Appleyard, M. M. (2007). Open innovation and strategy. *California Management Review, 50*(1), 57-76.

Clegg, C. W., Unsworth, K. L., Epitropaki, O., & Parker, G. (2002). Implicating trust in the innovation process. *Journal of Occupational and Organizational Psychology, 75*(4), 409-422.

Corbett, A. C. (2007). Learning asymmetries and the discovery of entrepreneurial opportunities. *Journal of Business Venturing, 22*, 97-118.

Davidsson, P., & Honig, B. (2003). The role of social and human capital among nascent entrepreneurs. *Journal of Business Venturing, 18*, 301-331.

Dodgson, M., Gann, D., & Salter, A. (2006). The role of technology in the shift towards open innovation: The case of Procter & Gamble. *R&D Management, 36*(3), 333-346.

Dudley, W. C., & Hubbard, R. G. (2004, November). *How capital markets enhance economic performance and facilitate job creation*. New York, NY: Global Markets Institute, Goldman Sachs.

Duxbury, S. (2012, January 6). North Face on $3B trek. *San Francisco Business Times*. Retrieved January 28, 2015, from http://www.bizjournals.com/sanfrancisco/print-edition/2012/01/06/north-face-on-3b-trek.html?page=all.

Easley, D., & Kleinberg, J. (2010). *Networks, crowds, and markets: Reasoning about a highly connected world*. Cambridge, England: Cambridge University Press.

Edgar, K. J., Buchanan, C. M., Debenham, J. S., Rundquist, P. A., Seiler, B. D., Shelton, M. C., & Tindall, D. (2001). Advances in cellulose ester performance and application. *Progress in Polymer Science, 26*(9), 1605-1688.

Evans, A., & Annunziata, M. (2012, November 26). *Industrial Internet: Pushing the boundaries of minds and machines*. Retrieved January 28, 2015, from http://www.ge.com/docs/chapters/Industrial_Internet.pdf.

Gaglio, C. M. (2004, Winter). The role of mental simulations and counterfactual thinking in the opportunity identification process. *Entrepreneurship Theory and Practice*, 533–552.

Gaglio, C. M., & Katz, J. A. (2001). The psychological basis of opportunity identification: Entrepreneurial alertness. *Small Business Economics, 16*, 95–111.

Galbraith, C. S., DeNoble, A. F., Ehrlich, S. B., & Kline, D. M. (2007). Can experts really assess future technology success? A neural network and Bayesian analysis of early stage technology proposals. *Journal of High Technology Management Research, 17*(2), 125–137.

Galbraith, C. S., Ehrlich, S. B., & DeNoble, A. F. (2006). Predicting technology success: Identifying key predictors and assessing expert evaluation for advanced technologies. *Journal of Technology Transfer, 31*(6), 673–684.

Gielnik, M. M., Krämer, A. C., Kappel, B., & Frese, M. (2014). Antecedents of business opportunity identification and innovation: Investigating the interplay of information processing and information acquisition. *Applied Psychology, 63*(2), 344–381.

Gruber, M. (2004). Marking in new ventures: Theory and empirical evidence. *Schmalenbach Business Review, 56*, 164–199.

Henderson, P. (2014). *You can kill an idea, but you can't kill an opportunity*. Hoboken, NJ: Wiley.

Hills, G. (1995). Opportunity recognition by successful entrepreneurs: A pilot study. In W. D. Bygrave, B. J. Bird, S. Birley, N. C. Churchill, M. Hay, R. H. Keeley, & W. E. Wetzel (Eds.), *Frontiers of entrepreneurship research*. Wellesley, MA: Babson College.

Hulbert, B., Brown, R. B., & Adams, S. (1997). Towards an understanding of "opportunity." *Marketing Education Review, 7*(3), 67–73.

Kirzner, I. (1979). *Perception, opportunity and profit*. Chicago, IL: University of Chicago Press.

Kirzner, I. M. (1997). Entrepreneurial discovery and the competitive market process: An Austrian approach. *Journal of Economic Literature, 35*, 60–85.

Knight, G. A., &Tamer Cavusgil, S. (1996). The born global firm: A challenge to traditional internationalization theory. *Journal of International Business Studies, 35*, 124–141.

Loise, V., & Stevens, A. J. (2010). The Bayh–Dole Act turns 30. *Science Translational Medicine, 2*(52), 52cm27–52cm27.

Long, W., & McMullan, W. E. (1984). Mapping the new venture opportunity identification process. In J. A. Hornaday, F. A. Tardley, J. A. Timmons, & K. H. Vesper (Eds.), *Frontiers of entrepreneurship research*. Wellesley, MA: Babson College.

Lucas, H. C., & Goh, J. M. (2009). Disruptive technology: How Kodak missed the digital photography revolution. *Journal of Strategic Information Systems, 18*(1), 46–55.

Merrilees, B. (2007). A theory of brand-led SME new venture development. *Qualitative Market Research: An International Journal, 10*(4), 403–415.

Moore, G. A. (1991). *Crossing the chasm: Marketing and selling high-tech products to mainstream customers*. HarperBusiness; Revised edition (August 2006). NY.

Osterwalder, A., Pigneur, Y., & Tucci, C. L. (2005). Clarifying business models: Origins, present, and future of the concept. *Communications of the Association for Information Systems, 16*(1), 1.

Palich, L. E., & Bagby, D. R. (1995). Using cognitive theory to explain entrepreneurial risk-taking: Challenging conventional wisdom. *Journal of Business Venturing, 10*, 425–438.

Parker, S. C. (2006). Learning about the unknown: How fast do entrepreneurs adjust their beliefs? *Journal of Business Venturing, 21*, 1–26.

Pinchot, G., & Pellman, R. (1999). *Intrapreneuring in action: A handbook for business innovation.* San Francisco, CA: Berrett-Koehler.

Puls, J., Wilson, S. A., & Hölter, D. (2011). Degradation of cellulose acetate-based materials: A review. *Journal of Polymers and the Environment, 19*(1), 152–165.

Quinn, J. B. (2000). Outsourcing Innovation: The New Engine of Growth. *MIT Sloan Management Review.* Summer 2000, *41*(4).

Ram, A. Tulsi. (1990). Archival preservation of photographic film—A perspective. *Polymer Degradation and Stability, 29*, 3–29.

Ramus, C. A. (2002). Encouraging innovative environmental actions: what companies and managers must do. *Journal of World Business, 37*, 151–164.

Rice, M., Kelley, D., Peters, L., & Colarelli O'Connor, G. (2001). Radical innovation: Triggering initiation of opportunity recognition and evaluation. *R&D Management, 31*(4), 409–420.

Roy, D., Semsarilar, M., Guthrie, J. T., & Perrier, S. (2009). Cellulose modification by polymer grafting: A review. *Chemical Society Reviews, 38*(7), 2046–2064.

Ryan, J. (2010). *A History of the Internet and the Digital Future.* Reaktion Books; Clerkenwell, UK.

Saxe, J. G. (n.d.). *The poems of John Godfrey Saxe.* Retrieved January 28, 2015, from https://archive.org/details/poemsjohngodfre02saxegoog.

Schumpeter, J. A. (1950). *Capitalism, socialism, and democracy* (3rd ed.). New York, NY: Harper.

Shane, S. (2003). *A general theory of entrepreneurship: The individual–opportunity nexus.* Northampton, MA: Elgar.

Shane, S., & Venkataraman, S. (2000). The promise of entrepreneurship as a field of research. *Academy of Management Review, 25*, 217–226.

Shepherd, D. A., & DeTienne, D. R. (2005). Prior knowledge, potential financial reward, and opportunity identification. *Entrepreneurship Theory and Practice, 29*, 91–112.

Short, J. C., Ketchen, D. J., Shook, C. L., & Ireland, R. D. (2009). The concept of "opportunity" in entrepreneurship research: Past accomplishments and future challenges. *Journal of Management, 36*, 40–65.

Simon, M., Houghton, S. M., & Aquino, K. (1999). Cognitive biases, risk perception and venture formation: How individuals decide to start companies. *Journal of Business Venturing, 15*, 113–134.

The North Face. (n.d.). *Elizabeth Hawker.* Retrieved January 28, 2015, from http://www.thenorthface.com/en_US/exploration/athletes/16-elizabeth-hawker.

The North Face. (n.d.). *FlashDry.* Retrieved January 28, 2015, from http://www.thenorthface.com/en_US/innovation/product-technology/flash-dry.

Timmons, J. A. (1994). Opportunity recognition: The search for higher potential ventures. In W. D. Bygrave (Ed.), *The portable MBA in entrepreneurship* (pp. 26–54). New York, NY: Wiley.

Truth: An Elephant in the Dark. (2012, October 13). Retrieved January 28, 2015, from https://anamaujood.wordpress.com/2012/10/13/truth-an-elephant-in-the-dark.

Tverksy, A., & Kahneman, D. (1974). Judgment under uncertainty: Heuristics and biases. *Science, 185,* 1124–1131.

Ucbasaran, D., Westhead, P., & Wright, M. (2008). Opportunity identification and pursuit: Does an entrepreneur's human capital matter? *Small Business Economics, 30,* 153–173.

Ucbasaran, D., Westhead, P., & Wright, M. (2009). The extent and nature of opportunity identification by experienced entrepreneurs. *Journal of Business Venturing, 24,* 99–115.

Ward, T. B. (2004). Cognition, creativity, and entrepreneurship. *Journal of Business Venturing, 19,* 173–188.

Weinstein, A. (1994). Market definition in technology-based industry: A comparative study of small versus non-small companies. *Journal of Small Business Management, 32*(4), 28.

West, J., & Lakhani, K. R. (2008). Getting clear about communities in open innovation. *Industry and Innovation, 15*(2), 223–231.

6

Better, Faster, Safer, and Cheaper

USAA Roof Inspections with Pole Cam

CLIFF ZINTGRAFF, MATT REEDY, SHANE OSBORNE, AND ROB PACHECO ■

INTRODUCTION

Many discussions of innovation processes can seem overly academic. Discussions often focus at a high level on broad data that suggests the success of various innovation practices. This chapter is a counterbalance to that trend. The focus is placed on a specific case of open innovation.

In the pursuit of their respective business and academic missions, the United Services Automobile Association (USAA) and the IC² Institute at the University of Texas at Austin (UT Austin) apply concepts of open innovation (Chesbrough, 2003) to the commercialization of products and services. At USAA, efforts have been underway for years that apply systematic innovation practices. The IC² Institute's core mission is the study and application of technology-driven economic development practices (IC² Institute, 2015).

In 2011, USAA and UT Austin formed a research and education partnership. The flagship program of the partnership is the UT Austin-led

USAA Innovator Certification (Reedy, Pacheco, Osborne, & Zintgraff, 2012). As part of one Innovator Certification cohort, a technology and business model emerged to address an unfolding business challenge around the quality, safe, and cost-effective inspection of roofs. This chapter describes how the technology was developed, and it describes how the innovation process was facilitated by the USAA Innovator Certification and the USAA–UT Austin partnership.

To place this open innovation example in context, the chapter begins with a brief overview of USAA and its mission. Then, the chapter examines the collaboration between USAA and the IC^2 Institute at UT Austin. It explores how USAA and UT Austin collaborated to research the proposed technology and business model for inspections of roofs by insurance adjusters. The chapter describes how USAA followed through to achieve a successful innovation. The chapter concludes by suggesting lessons that can be drawn from the current case and collaboration.

ABOUT USAA

USAA was founded by US Army officers in 1922. The founding of the association is summarized as follows (USAA, 2015a):

> In 1922, when 25 Army officers met in San Antonio, Texas, and decided to insure each other's vehicles, they could not have imagined that their tiny organization would one day serve millions of members and become one of the only fully integrated financial services organizations in America.

USAA's founders infused the organization with their shared military values of "service, loyalty, honesty and integrity" (USAA, 2015a, Corporate Overview, para. 2). USAA's mission is to "facilitate the financial security of its members, associates, and their families through provision of a full range of highly competitive financial products and services; in so doing,

USAA seeks to be the provider of choice for the military community" (USAA, 2015a, Our Mission, para. 1).

USAA is structured as a member-owned inter-insurance reciprocal exchange (USAA, 2015b) and thus is not subject to many of the pressures and constraints of typical private and public corporations. Eligibility for USAA membership is a privilege earned by those in uniform that can be handed down to their children. Membership is open to active, retired, and honorably separated officers and enlisted personnel of the US military and their families and also to officer candidates in commissioning programs. USAA's products include auto, property, and life insurance; banking products such as checking, credit cards, and auto and home loans; investment products such as mutual funds, brokerage accounts, and managed portfolios; and special discounts on shopping for retail products and services (USAA, 2015a).

USAA highlights, for differentiation from competitors, its more than 10 million "passionate and loyal" members and its more than 27,000 world-class employees who are personally committed to delivering excellent service and valuable advice (USAA, 2015c). In addition, USAA maintains high financial strength ratings from three of the major rating agencies: USAA is one of two US property and casualty companies with the highest ratings from A.M. Best and Moody's Investors Service (USAA, 2015c). The company states, "When people join USAA, they join generations of military families who have depended on the company to provide superior products and services in an atmosphere of financial strength" (USAA, 2015a, Stronger Together, para. 1).

INNOVATION AT USAA

USAA believes that innovation starts with its corporate culture. Innovation Communities for the Enterprise (ICE), an online community, allows employees to submit innovative solutions, participate in specific enterprise challenges, collaborate with peers, and participate in solution development. USAA's lines of business (e.g., property and casualty, banking, and investing), USAA's information technology resources, and other support resources can then join to pilot new capabilities in one of the in-house labs designed to test prototypes in a variety of scenarios. More than 95% of USAA's employees participate in the innovation community by contributing more than 10,000 ideas each year to help simplify and improve the lives of USAA's members.

USAA has demonstrated leadership in mobile financial services, evidenced by recent innovations. USAA was the first financial institution to offer mobile check deposits without requiring submission of the original check (Patterson, 2007; Stanier, 2011). With the more recent virtual mobile assistant, members can use voice commands to navigate and complete nearly 200 banking transactions on the USAA mobile app. The same mobile app supports the purchase of life insurance online. It also makes it easy and convenient to report claims. Using an iPhone, iPad, or Android device, members can make a claims report, upload pictures, schedule appraisals, make rental car reservations, and check claim status. An accident animation feature lets members explain their accident via animation and narration (USAA, 2015d).

Other recent additions to USAA's mobile app include person-to-person payments and mobile investing. With person-to-person payments, members can pay almost anyone from the mobile app. The new mobile investing platform allows members to "trade stocks; [trade] mutual funds and CDs; access real-time quotes and market news; and deposit checks directly into eligible investment accounts" (USAA, 2015d, para. 8).

Beyond the ICE platform, USAA's innovation team employs a number of research engineers that continually identify and evaluate leading-edge technologies and business models with the potential to benefit USAA members. This *revolutionary innovation* organization focuses on developing ideas that fill a *real* need in a *new* way that provides a *remarkable* experience. Frequent idea-generation sessions lead to ideas being contributed to an incubation queue in which prototypes, proofs of concept, and video vignettes are created and for which vendor evaluations are conducted. By combining the best of USAA's internal resources with the best available external resources, USAA pursues a strategy to be the innovation leader in the financial services industry. Ideas that are successfully incubated are directed to delivery teams within the company. The delivery teams deploy agile methodologies to deliver these innovations to USAA's members as quickly as possible.

THE USAA–UT AUSTIN COLLABORATION

The USAA–UT Austin collaboration builds on the organizations' respective efforts to date. The idea of a research partnership began in 2010. While meeting at a community event, the CEO of USAA and the chancellor of The University of Texas System discussed the possibility of such a program. Soon, a memo was exchanged between the CEO and the chancellor agreeing in principle to explore a sponsored research program.

At that moment, the discussion expanded to include the IC^2 Institute. Founded in 1977 by Dr. George Kozmetsky, the IC^2 Institute resides in the portfolio of the UT Austin Vice President for Research (IC^2 Institute, 2015). The IC^2 Institute's efforts focus simultaneously on innovation research activities and the application of that research in practice. This approach embodies the philosophy Dr. Kozmetsky had for the IC^2 Institute, where research is translated into practice. For this reason, the IC^2 Institute is often referred to as a "'think and do' tank" (para. 1) by its staff. Dr. Kozmetsky, a recipient of the Presidential Medal of Freedom for Technology from President Bill Clinton, worked on broad, unstructured problems, and he believed that such efforts should integrate research and practice. Fields of study such as grounded theory (Glaser & Strauss, 2009) and design-based research (Barab & Squire, 2004) influence, and are evoked by, the work of the IC^2 Institute.

Today, the IC^2 Institute is home to the Austin Technology Incubator (ATI), the Global Commercialization Group, the Bureau for Business Research, the IC^2 Institute Global Fellows Network, and the Southwest Node for the National Science Foundation ICorps program for technology transfer (IC^2 Institute, 2015). Its methodologies have been applied with the more than 200 company graduates of ATI, with government and institutional partners in more than 20 countries, in entrepreneurial education programs, and in domestic programs such as the one that is the subject of this chapter.

In a university–industry collaboration such as the one between USAA and UT Austin, what might be the focus? Early in the process, two large contingents from USAA visited the IC² Institute. Among the topics discussed as potential focus areas were faculty-sponsored research, student projects, student internships, technology portfolio development, and technology transfer practices. Some of these topics became part of the program; the focus area selected was *innovation education and training*. Specifically, the program would focus on assessment of technology ideas and related business models for technology development. To support this focus, the IC² Institute recommended the Quicklook[1] method (Cornwell, 1998) as the core topic of instruction.

Used in IC² Institute programs for approximately 20 years, the Quicklook method teaches participants how to rapidly perform a market-based, primary research-focused analysis of a technology idea. The method was born from the need to focus limited innovation resources on the most promising new technologies. This kind of decision-making has historically been ad hoc in nature, overly focused on single issues, poor at recognizing early challenges and barriers, and/or disconnected from market realities as experienced by stakeholders and potential customers. Åstebro and Koehler (2007) referred to commercialization forecasts as frequently intuitive and miscalibrated relative to commercial potential.

The Quicklook method promotes good decision-making by engaging market stakeholders early and by leveraging the networks of those participants. Its process leads to market insights, identification of first market segments, and potential go-to-market strategies. The method is amenable to use with live cases. Framing education and training in live cases supports two program goals: (1) making an immediate impact on business goals and (2) improving retention of material and the ability to apply the material and method in context. More details on the Quicklook method, including background on its original development at the NASA Mid-Continent Technology Transfer Center, can be found in Cornwell (1998).

It was decided that the first activity of the USAA–UT Austin partnership would be a 25-student cohort in a course designed around the Quicklook method. The course would be called the USAA Innovator Certification. The 12-day, six-session course would lead to a UT Austin Certificate of

Completion for each USAA staff member completing the course. In the first cohort, four teams would analyze one technology idea per team, with the ideas drawn from existing USAA innovation processes. Each team would produce a Quicklook presentation that made a *yes, yes–if,* or *no* recommendation, with that recommendation presented to a joint, executive-level panel. The teams would be judged on the degree to which their recommendation was evidence-based and grounded in knowledge of their market and its stakeholders. Primary research would be the main mechanism for gathering of market insights. Of the four teams in the first cohort, one was researching how to improve the process of roof inspections. That problem is the focus of the case study in this chapter.

THE PROBLEM

One of the keys to providing exceptional value and service to the insurance consumer is completing claims and repairs quickly and cost-effectively. One of the largest expenses involved with homeowner's insurance claims is evaluating the damage to roofs following a weather event. Traditionally, these inspections have been done by climbing on the roof to visually assess damage, take photographs, and perform measurements necessary to create the estimate to repair. Performing this task both safely and quickly presents a huge challenge for insurance companies and their employees because of the inherent danger of walking on roofs and because of potentially difficult weather conditions that can hamper the work of adjusters.

With a growing membership and increasing safety regulation, USAA believed that something needed to be done to contain expenses and increase the safety of employees and anyone else performing this hazardous task.

FROM UNMANNED AERIAL VEHICLES TO POLE CAM

For the challenges associated with roof inspections, one possible solution is the use of unmanned aerial vehicles (UAVs). This was the original technology solution presented to the roof inspection team in the Innovator Certification's first cohort. In 2009 and 2010, USAA Innovation and the Property and Casualty staff noticed the advances in UAV technology, including the capability to remotely capture digital imagery and video. This inspired research to determine whether remote digital imagery capture could deliver equal or better results than the current physical inspection process, relative to accuracy and quality.

An innovation team began testing selected UAVs. The team invited a demonstration of a state-of-the-art, $17,000 UAV. It was operated with connectivity to a laptop computer. The UAV lost stability and connectivity in a 12-mile-per-hour wind. In addition to concerns with performance in the air, advancement was needed in ease of operation, battery life, and cost. It became apparent that UAV technology was not currently ready for the roof inspection challenge.

However, the technology and continuing advancement in digital cameras, video equipment, and Wi-Fi connectivity were very encouraging. One USAA employee had done a home experiment using a digital camera attached to a broomstick with duct tape. The employee had taken pictures of his roof from the safety of the ground, achieving excellent resolution and detail. This simple technology became an alternative direction available to the roof inspection team, and it was a potential subject for a Quicklook assessment. The team discussed the drivers and restrainers of both options on the first day of class and then chose to work on perfecting remote digital imagery using the simpler, pole-based solution. If this process was perfected, the majority of the process would be reusable if and when other methods for getting above the roofline became a reality. The Pole Cam experiment was born.

VETTING THE IDEA

The team immediately embraced the theory that high-quality images could be captured from a camera extended in the air on a pole. However, there was very little technology available to test the theory. The team had determined early on that the system had to be wireless because adjusters must navigate all kinds of different terrain and obstacles in homeowners' yards—"anything that could get caught by a yard gnome would not be safe or practical" (Roof Inspection Team, 2011, personal communication). The pole also had to extend high enough to easily inspect two-story homes, and the apparatus had to be both lightweight and rigid in order for an adjuster of any size to be able to handle it. That requirement eliminated fiberglass, which is neither lightweight nor rigid. It had to be nonconductive, which eliminated heavy aluminum poles. To complicate matters further, no manufacturers were building cameras that could be controlled remotely from a tablet or phone with a built-in viewfinder. The team members commented that they felt like "Howard Stark—Iron Man's dad—we had a great idea, but the technology was not yet invented to bring it to life" (Roof Inspection Team, 2011, personal communication).

Slowly but surely, the team found pieces and parts that enabled primitive testing. The team found a pole from England that was made of carbon fiber, meeting three of the four needs: sufficient length, lightweight, and rigid. The pole was conductive, but this was categorized as a long-term problem, and the pole was adopted for testing purposes. The team found a wireless backup camera that provided a low-definition picture of the roof, and then it added a GoPro[2] camera underneath to take pictures every 5 seconds. During tests, many pictures had to be taken to obtain the ones needed, but testing proved it was possible to see damage on shingles using a high-definition camera. Figure 6.1 includes a picture of the first Pole Cam prototype.

FIELD TRIALS: THE GOOD, THE BAD, AND THE UGLY

Once the team had a field-worthy prototype, field time was scheduled to test the product. One of the first adjusters to test the prototype previously had fallen from a roof, breaking an arm and a leg and spending 6 months out of work. His initial feedback strongly encouraged the team. He stated that he could see himself doing the job he loved for many more years because he would not have to climb on dangerous roofs anymore.

As the product was refined, a Wi-Fi-enabled camera was identified, allowing control of the camera with a tablet. This was a breakthrough: The user could take only the pictures actually wanted. Based on the positive feedback of the test adjusters, approval was secured from stakeholders to expand the pilot to adjusters around the country. However, prior to that step, there was a required safety and security review. USAA values employees and their safety. Nothing would be allowed in an employee's hands that had not been vetted thoroughly through the safety department.

The safety department was impressed with the fact that the poles were lightweight, especially compared to the ladders that adjusters used to inspect roofs. However, the department was concerned about electrical and lightning risk. To satisfactorily mitigate this risk, the roof inspection team worked with the pole manufacturer to build the same poles but with the outside wrapped in fiberglass. The result was a pole that insulated the adjuster from shock should the pole touch anything electrified. A physical test of the poles was performed at USAA's airplane hangar. The company's corporate jets are tested for protection from lighting strikes, and the responsible group had the equipment necessary to test the pole. The pole passed. The team joked that "no one felt anything!" (Roof Inspection Team, 2011, personal communication)—although to be absolutely clear, that test was performed using professional gear following safety protocols.

Next, it was necessary to determine where a 30-foot pole would be stored in the adjuster's car. The safety department insisted that the pole be stored in the trunk, co-located with the ladder, fire extinguishers, and other gear that an adjuster carries. This presented two problems: space and the potential for damage to the pole or camera from hard contact with the other items. It was necessary to determine the maximum length of pole and carrying case that would fit into the trunk of a company-standard Ford Fusion loaded with all an adjuster's equipment. Through personal experience, members of the roof inspection team knew that rifle cases are effective at protecting long, skinny, metal-type items. A member of the roof inspection team visited a sporting goods store and had the manager escort him to the car with five rifle cases. A case was selected that was a good fit for the Pole Cam.

The pole manufacturer later modified the number of sections in the pole to better accommodate the height and collapsed length requirements of the carrying case.

Note that selecting a rifle case as the carrying case for the pole created some new, unanticipated process challenges. As one might imagine, adjusters carrying camouflage rifle cases in their cars, and to members' homes, might cause some concerns. To mitigate this concern, the cases were painted white by USAA's paint shop, and they were adorned with USAA's logo, all to make them look less ominous. However, this did not satisfy requirements for carrying the Pole Cam through USAA's campus. Therefore, for all Pole Cam presentations outside the Innovation Lab, a security escort was required to carry the case to its destination. The team received many strange looks, but at least they were innovating.

SELLING THE BUSINESS: CONVINCING STAKEHOLDERS TO DRAMATICALLY CHANGE THEIR PROVEN APPROACH TO ROOF INSPECTIONS

The first response received from line-of-business stakeholders when the idea of Pole Cam was unveiled was to ask what happened to the UAV. After that, the stakeholders expressed doubts that a system such as Pole Cam could capture the data necessary to determine roof damage. In hindsight, they had every right to be skeptical. Adjusters on a roof can touch and bend shingles to get a feel for damage. The roof inspection team was proposing to remove this technique from its assessment for most inspections.

However, two issues would be paramount in establishing the value of Pole Cam. First, it was clear to all involved that Pole Cam provided substantial safety benefits for employees. This benefit was decided quickly. Second, the Wi-Fi camera technology that emerged created productivity benefits that convinced stakeholders that Pole Cam would benefit the work of adjusters. An unproductive Pole Cam would simply drive adjusters back onto roofs, but a more productive Pole Cam would create a virtuous cycle between safety improvement and better productivity. The final

Figure 6.1 Pictures of Pole Cam. Used with permission from the USAA.

Wi-Fi camera technology featured a tablet-controlled camera with 21× optical zoom. An adjuster could stand on the street and zoom in all the way to the granules on the shingles. After 10 months of field testing, the line-of-business stakeholders made the decision to roll out Pole Cam to the entire adjuster team. Figure 6.1 includes pictures of different versions of Pole Cam.

MANUFACTURING AND TRAINING

The decision to roll out Pole Cam created demand for approximately 200 field units to be delivered throughout the United States. During initial stages and the pilot, a roof inspection team member had built all of the components necessary to hold the camera and the tablet to the pole. The thought of individually producing 200 components that would survive the real world of property inspections was daunting. Nevertheless, all other components had been approved by line-of-business stakeholders, by security, and as necessary by USAA's branding department. Also, the uniqueness of the connector hardware and related components made it difficult to find a partner or machine shop willing to produce those components. To date, they had been fabricated in-house, with raw materials and piece parts procured.

Therefore, during the course of approximately 2 months, a member of the roof inspection team built, on nights and weekends, 200 individual camera mounts and iPad mounts. Also built was a creeper tool that used add-on hardware and wheels to enable inspection around chimneys and other obstacles on a roof. This team member's shop at home had been overrun with "all manner of strange-looking contraptions" (Roof Inspection Team Member, 2012, personal communication). All parts were relocated to the USAA maintenance department, and the team proceeded to assemble 200 units and prepare them for shipping to adjusters.

Before Pole Cam could be shipped to adjusters, it was necessary to determine how to train adjusters on proper use of the device. Some team members wanted to visit with all the adjusters, whereas others wanted to produce a video to send with each Pole Cam to all the adjusters. Cost concerns led to selection of the video option. On the video, one of the pilot adjusters showed how to use Pole Cam and discussed strategies he had learned to get the most use out of Pole Cam. Another team member demonstrated connecting the camera to the tablet computer and related functionality.

ROLE OF THE QUICKLOOK PROCESS AND THE USAA–UT AUSTIN COLLABORATION

As previously noted, the USAA Innovator Certification course is organized around the Quicklook process (Cornwell, 1998). During the process, analysts create a short (<1 page) description of a technology-based idea, including a layman's description, description of benefits, support through as many quantitative measures/estimates as possible, comparison to competing solutions, and evidence of progress in technology development. Then, the description is shared with stakeholders, including potential customers (who would pay for the technology), end users (who would use the technology), delivery partners, scientific experts, and investors (anyone who might invest money or materials to develop the technology). A short interview is conducted soliciting open-ended feedback. Through six course sessions each lasting 2 days, participants research their idea, construct a market-focused presentation and recommendation (*yes, conditional yes*, or *no*), and earn a Certificate of Completion from UT Austin.

A crucial part of each interview is a request for additional contacts—it is through this *degrees-of-separation* process that deep experts are identified who can rapidly and expertly inform the research. Continual refinement of the technology description leads, for good technology ideas, to a highly refined technology description that forms the core of a Quicklook report and/or presentation. As a heuristic, 8–12 interviews are typically needed to reach a confident conclusion about a technology idea; however, in the experience of IC^2 Institute trainers, more interviews are typically completed in training settings such as the USAA Innovator Certification. Figure 6.2 presents a simplified illustration of the Quicklook process.

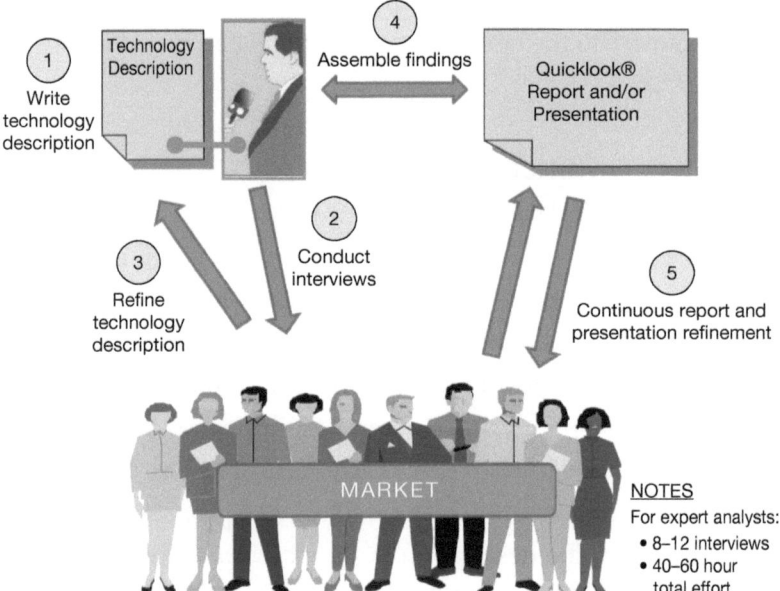

Figure 6.2 Quicklook process simplified illustration.

The Quicklook process and the USAA Innovator Certification course played a substantial role in the development of an idea that was underway, but only lightly researched, at the start of the first Innovator Certification cohort. The emerging problems with roof inspection had already been recognized, and the idea of using UAVs for inspection had been pursued prior to the course. As the class began, it was becoming clear that although the use of a UAV might eventually provide a superior method for roof inspections, UAV technology needed to be further investigated and developed for roof inspection use cases. Although the team could have simply accepted and reported this result, it instead embraced a core tenet of the Quicklook process, which is to always return to the problem being solved and consider alternatives.

The roof inspection team credits a member of USAA Innovation's research team with the idea of using a pole for inspections. Facilitated by the process and course sessions, the team set aside the UAV idea, substituted the pole camera idea, wrote a technology description, and began conducting stakeholder interviews. By the end of the course, through its primary research (interviews) and secondary research, the team had constructed its first Pole Cam prototype; demonstrated usage; collected high-quality roof pictures from early tests; and developed a story of the business benefits based on safety, productivity, and technical feasibility. Along the way, the roof inspection technology idea was refined; a new approach was considered, researched, and recommended; and a working prototype was developed with the important creeper add-on to capture pictures behind chimneys and other obstacles.

By the end of the course, the team was able to make a confident recommendation about a workable technology, how that technology would fit USAA's roof inspection business model, and how it would increase the safety and productivity of adjusters. Although the product would undergo significant revision after the course (better camera, better remote control, and improved pole apparatus), the team built the foundation for success during the joint USAA–UT Austin efforts during the Innovator Certification course.

POLE CAM: AN OPEN INNOVATION SUCCESS

To date, Pole Cam is considered a major success. Safety has been improved, and increased efficiencies have been measured and shared with stakeholders inside USAA. Adjusters have widely and enthusiastically adopted Pole Cam. Along the way, one can see the concepts of open innovation at work. Within the company, a robust idea system and associated culture led to the original idea of using UAVs for inspection. A search for best-in-class vendors led to understanding what innovations existed, and which ones did not exist, to support the idea. Persistence and a systematic approach led to finding new approaches to the problem and also to simpler technologies (in this case, the expanding and portable pole) that could address the original problem. Broad inclusion of stakeholders and flexible use of the energy and creativity of employees were key to seeing the process through. Techniques for research, and general encouragement of an innovation culture, were provided by the Quicklook process and by the ongoing innovation partnership with USAA and the IC^2 Institute at UT Austin.

A proper note on which to close this discussion regards a later and inspired use of Pole Cam. In the summer of 2012, USAA adjusted losses on homes damaged by the Waldo Canyon fire near Colorado Springs, Colorado. A USAA adjuster inspected fire-gutted homes, without risking going inside, by extending a Pole Cam horizontally into the home to take pictures. This kind of innovative thinking applied the Pole Cam technology in a way never anticipated. This may become the inspiration for new applications or for new products that improve employee safety, productivity, and service to USAA members.

THE PARTNERSHIP CONTINUES

The USAA–UT Austin research partnership continues to thrive. The USAA Innovator Certification is starting its sixth year and its eleventh cohort. Program accomplishments include immediate business impact on projects, a significant number of USAA staff certified in the Quicklook process, a newly certified group of in-house Quicklook mentors, and go-to-market strategies for a variety of innovations. The Quicklook method is integrated at key points in the larger USAA development process. Not to be overlooked are the *no* recommendations that have freed up resources for other efforts and the *yes–if* recommendations that have driven important course corrections. Outside the USAA Innovator Certification, there have been three faculty-supported research projects, six graduate student team projects, and other faculty visits and speaking engagements designed to promote creative, evidence-based, market-driven innovation as a culture change that should permeate the organization.

In the final analysis, the Quicklook process, primary research tools, methods, and USAA Innovator Certification program are simply tools in the hands of skilled professionals. What is provided is a systematic method for driving innovation over time, adapted for local context. The Quicklook process is built on habits of mind, such as recognition of inconsistencies, self-awareness of when understanding exists and when it does not, and strategies for recognizing important information and applying it for maximum benefit. Stated simply, the process is built on precepts of smart thinking (Markman, 2012). Innovation must be systematic to make sustained impact (Cooper, 1990; Cooper, Edgett, & Kleinschmidt, 2002), but it is also a disciplined and creative process, not unlike many other human endeavors, and it comes down to the individual actions of people and the choices they make. The development of greater human capacity for innovation is the greatest long-term impact of the USAA–UT Austin partnership.

POSTSCRIPT

As this chapter went to press, there was late-breaking news: Pole Cam was commercialized. Specifically, the Pole Cam technology was licensed by USAA to a third party that will apply it to other applications and industries. The details of the relationship are private, but the commercialization success is an important marker for USAA. In solving an internal problem, USAA created a technology that can address other problems and, in doing so, even better justify the original investment. USAA's ongoing program to find innovative solutions to problems will benefit from both the experience of Pole Cam and the new revenue earned from licensing Pole Cam.

Notes

1. Quicklook is a registered trademark of The University of Texas at Austin.
2. GoPro is a registered trademark of GoPro, Inc.

References

Åstebro, T., & Koehler, D. J. (2007). Calibration accuracy of a judgmental process that predicts the commercial success of new product ideas. *Journal of Behavioral Decision Making, 20*(4), 381–403.

Barab, S., & Squire, K. (2004). Design-based research: Putting a stake in the ground. *Journal of the Learning Sciences, 13*(1), 1–14.

Chesbrough, H. W. (2003). *Open innovation: The new imperative for creating and profiting from technology.* Boston, MA: Harvard Business School Publishing.

Cooper, R. G. (1990). New products—What distinguishes the winners. *Research Technology Management, 33*(6), 27–31.

Cooper, R. G., Edgett, S. J., & Kleinschmidt, E. J. (2002). Optimizing the Stage-Gate process: What best-practice companies are doing—Part I. *Research Technology Management, 45*(5), 21–27.

Cornwell, B. (1998). "Quicklook" commercialization assessments. *Innovation: Management, Policy & Practice, 1*(1), 7–9.

Glaser, B. G., & Strauss, A. L. (2009). *The discovery of grounded theory: Strategies for qualitative research.* New Brunswick, NJ: Transaction Publishers.

IC² Institute. (2015, January). *About the IC² Institute.* Retrieved January 12, 2015, from http://ic2.utexas.edu/about.

Markman, A. (2012). *Smart thinking: Three essential keys to solve problems, innovate, and get things done.* New York, NY: Penguin.

Patterson, S. (2007, January). *USAA Deposit@Home—Another WOW moment for net banking.* Retrieved January 12, 2015, from http://www.creditunions.com/articles/usaa-deposit-home-another-wow-moment-for-net-banking/#ixzz3OeRPioX3.

Reedy, M., Pacheco, R., Osborne, S., & Zintgraff, C. (2012, November). *Open innovation case study: Intrapreneurship: From drone to Polecam.* Paper presented at OI: The Journey from Ideation to Innovation, Austin, Texas.

Stanier, J. (2011). An interview with Greg Schwartz. *XRDS: Crossroads, the ACM Magazine for Students, 17*(3), 28–30.

USAA. (2015a, January). *Corporate overview.* Retrieved January 12, 2015, from https://www.usaa.com/inet/pages/about_usaa_corporate_overview_main.

USAA. (2015b, January). *Enterprise: Annual report*. Retrieved January 12, 2015, from https://www.usaa.com/inet/pages/g_old_Enterprise_Annual_Report_index.

USAA. (2015c, January). *Financial strength*. Retrieved January 12, 2015, from https://www.usaa.com/inet/pages/about_usaa_corporate_overview_financial_strength.

USAA. (2015d, January). *Bringing simplicity through innovation*. Retrieved January 12, 2015, from https://content.usaa.com/mcontent/static_assets/Media/USAA_Fact_Sheet_Innovation.pdf?cacheid=3999054578_p.

7

An Open Invitation to Open Innovation

Guidelines for the Leadership of Open Innovation Processes

DIANA RUS, BARBARA WISSE, AND ERIC F. RIETZSCHEL ■

Innovation is crucial to how organizations create value for themselves, secure a position of competitive advantage, and attain market leadership (Christensen, 1997; Davis & Eisenhardt, 2011). For instance, in a study by Bain & Company among 1,208 global executives, 74% identified innovation as more important than cost reductions for the long-term success of their companies (Rigby & Bilodeau, 2013). However, in increasingly complex, dynamic industries with frequently changing and highly dispersed resources, it is unlikely that any single organization can consistently deliver breakthrough innovations on its own. Instead, more open and collaborative approaches to innovation are needed to enhance internal innovation capabilities and to secure a position of competitive advantage (Chesbrough, 2003, 2011; Chesbrough, Vanhaverbeke, & West, 2006; Dahlander & Gann, 2010).

During the past decade, the academic and popular literature on open innovation (OI) has increased exponentially (for reviews, see Chesbrough & Bogers, 2014; Huizingh, 2011; van der Vrande, Vanhaverbeke, & Gassmann, 2010). Moreover, companies such as Procter & Gamble, Hewlett-Packard, Intel, IBM, and Genzyme are often credited with having strongly profited from adopting the OI paradigm, by collaborating with external partners (Whelan, Parise, de Valk, & Aalbers, 2011). Fueled by the successes of these trailblazing companies, an increasing number of organizations of different sizes and across industries have followed suit in adopting the OI paradigm in the hope of reaping the same rewards. However, many of them are finding it difficult to replicate these successes and fully realize the benefits associated with OI (Mortara & Minshall, 2011; Slowinski & Sagal, 2010). Indeed, evidence suggests that returns from OI vary substantially among companies, with a large number of firms reporting failed OI projects (Laursen & Salter, 2006). However, little is known about why some firms are capable of profiting from OI, whereas others are not.

In this chapter, we argue that one of the main reasons why some organizations fail to fully capitalize on the promise of OI is their neglect of the intraorganizational, micro-level foundations of OI implementation. That is, we contend that some organizations that shift from a more closed to a more open innovation model fail to address the question of how to organize and manage OI internally. We therefore advocate taking an intraorganizational perspective by zooming in on leader behaviors and practices that, we believe, are crucial for successful OI implementation. In the following, we first provide more specific reasons as to why we think that a micro-level perspective in general, and a focus on leader behaviors in particular, is valuable in helping us better understand and manage the OI implementation process. We then show the value of taking a micro-level perspective by discussing how leaders can play a vital role in the OI implementation process, and we provide some much-needed practical advice to executives mired in the fog of embedding OI within their organizations.

THE ADDED VALUE OF TAKING A MICRO-LEVEL PERSPECTIVE ON OPEN INNOVATION

To date, most of the literature on OI has concentrated on identifying the macro-level organizational practices firms use to leverage external sources of knowledge and capture value from collaboration by boosting internal innovation (Salter, Criscuolo, & Ter Wal, 2014; van der Vrande et al., 2010). In contrast, our understanding of the micro-level foundations of OI adoption—the internal processes and dynamics of implementation, especially as they relate to the employees and leaders involved—is lagging behind (da Mota Pedrosa, Välling, & Boyd, 2013; Mortara & Minshall, 2014). Indeed, the individual and team levels of analysis are the most underinvestigated areas in the OI domain (Chesbrough & Bogers, 2014). However, there are several reasons why taking a micro-level perspective on OI in general, and a leadership perspective in particular, is essential and in need of more attention.

First, a successful shift from a paradigm of closed innovation to one of open innovation needs to be underpinned by foundations on two levels: (1) the *macro-level foundations* of (re)structuring units, workflows, and alliances involving potentially risky collaborations with external partners and (2) the *micro-level foundations* of instigating and leading a significant organizational change process (Mortara & Minshall, 2014; Salter et al., 2014). Thus, practitioners always need to balance two distinct sets of concerns: one focused on organizational structures and linkages with external collaborators and one focused on the process of managing implementation internally.

Second, it is precisely the (understudied) micro-level processes of change management that seem to make successful OI implementation so difficult. Many companies have found their OI efforts stalled in large part due to internal barriers related to the human side of OI (Salter et al., 2014). For instance, a recent survey among executives of 125 European and American large firms found that the major challenges in managing OI are within the firm, with executives rating the change process from closed to open innovation as the most difficult task (Chesbrough & Brunswicker, 2013). Although the role of management has been touted as crucial in overcoming the internal challenges of OI implementation (Chiaroni, Chiesa, & Frattini, 2011; Giannopolou, Yström, & Ollila, 2011; Mortara, Napp, Slacik, & Minshall, 2009), practical guidelines based on systematic empirical research are largely missing.

Third, even those processes that take place at the macro level often find their foundations at the micro level. The shift from closed to open innovation inevitably requires profound changes in terms of internal processes and structures. However, this always involves changes in the attitudes, mindsets, and behaviors of organizations' most "basic" constituents, namely the individuals within them. It is individuals that make decisions and act upon them, not organizations (Foss, Husted, & Michailova, 2010). Given that individuals are the driving force behind all organizational processes (Senge, 1990), a micro-level perspective is essential to our understanding of OI processes.

Fourth, a micro-level perspective on leader behaviors may help organizations capture concrete opportunities to successfully manage OI processes because out of all organizational members, leaders exert disproportionate amounts of influence on employee attitudes, motivation, behavior, and performance (for reviews, see Boal & Hooijberg, 2001; Yukl, 2008). Moreover, leaders have been identified as being crucial in sparking, driving, and sustaining organizational culture change (Kanter, 1983; Schein, 1993, 2004). However, although leaders' critical role in successful OI adoption has often been acknowledged (Elmquist, Fredberg, & Ollila, 2009; Mortara & Minshall, 2014), there is a relative paucity of research or practical guidelines devoted to understanding their role in successfully managing OI.

The ensuing sections of this chapter are therefore dedicated to demonstrating how a micro-level perspective can be useful in facilitating the successful implementation of OI programs. We focus on the role of leadership in particular because, as stated previously, leaders can potentially influence the OI implementation process to a great extent. By drawing on our own experience with the implementation of OI programs as well as on empirical evidence from the leadership, organizational change, and open innovation literatures, we provide several guidelines that may increase leaders' chances of successful internal OI adoption. Note that we do not claim to be exhaustive; that would take several volumes instead of one chapter. We simply want to show that a micro-level perspective can help organizations seize and capitalize on opportunities to manage OI processes successfully. To this end, the remainder of this chapter is dedicated to answering the following question: "How can leaders contribute to the successful embedding of OI within their organizations?"

THE ROLE PLAYED BY LEADERS IN THE INTERNAL OPEN INNOVATION IMPLEMENTATION PROCESS

We identify four critical aspects of OI implementation in which leaders can play a vital role in ensuring success. All four aspects must be managed carefully because they are central to making or breaking employee motivation and the ability to engage in OI. Specifically, we focus on the role that leaders play in getting people on board, reassuring employees that OI is here to stay, reshaping the larger definition of success within the firm, and enabling internal capability building. Importantly, for each of these four aspects of OI implementation, we outline two possible practical strategies (based on empirical findings) that leaders could bring to bear to overcome impediments related to issues of employee motivation and ability.

Get People on Board

> Provide opportunities for people to get on board. As with any fundamental shift, it will take time for people to fully digest the change that open innovation represents and its implications on the business and their specific role.
> —TODD BOONE, *Director of Market Development at Psion*[1]

Once a decision has been made by top management to make OI a strategic priority, it is imperative that the rest of the organization follows suit and embraces a shared view of open innovation. This much is obvious. However, it is often less obvious how this can best be achieved. Next, we outline two main strategies that leaders can employ to promote a shared view of open innovation within the organization: craft and communicate a compelling vision and promote shared ownership.

Craft and Communicate a Compelling Vision

One of the most effective ways to get people on board is to inspire them by advancing a simple, compelling vision of a desirable future (Conger & Kanungo, 1998; Shamir, Arthur, & House, 1994). Visions provide abstract guidelines that help direct the behaviors of organizational members by providing an ideal sketch of what the organization may look like in the future. Although goals and visions are similar, they are not the same. Visions are more abstract and global than goals, and unlike goals, they are not necessarily meant to be fully achieved (Kirkpatrick & Locke, 1996). A clear vision of OI can, for instance, incorporate what the organization tries to become by opening up the innovation process (e.g., the aim of Philips Research to "become a leader in health and well-being"[2]).

Organizational members can use the vision as a sort of standard against which they can hold their behaviors, thoughts, and feelings and assess their appropriateness (e.g., "Does this decision to bring in external knowledge help us become a leader in health and well-being?"). As such, a vision that incorporates OI promotes a set of desirable attitudes, values, and beliefs (e.g., collaboration, openness, and borrowing with pride). Importantly, a vision that creates a common sense of purpose and meaning around a set of collective goals by stressing a shared, organizational identity is more likely to inspire and motivate a wide array of internal stakeholders (Awamleh & Gardner, 1999; Stam, van Knippenberg, & Wisse, 2010). Focusing on collective goals and promoting a shared identity at the organizational level that supersedes individual, team, or business unit identities is especially important when shifting to OI, given the increased need for collaboration between people in different parts of the organization that may traditionally have pursued competing goals.

The success of a vision is contingent on how well it is communicated (Awamleh & Gardner, 1999), and this, in turn, is contingent on several leader behaviors. First, visions are more persuasive when they are communicated with the use of colorful language that appeals to people's imagination and emotions. Skillful leaders address their audience directly and incorporate rhetoric devices, vivid imagery, metaphors, analogies, and symbols into their communication (Emrich, Brower, Feldman, & Garland, 2001). For instance, one vivid metaphor that has been effectively used in different contexts (e.g., the pharmaceutical industry and the retail industry) to depict the need for communication, coordination, and collaboration is that of a NASCAR race crew (Roberto & Levesque, 2005; Sutton & Rao, 2014). In addition, successful vision communicators use slogans, anecdotes, and tell compelling stories of past successes (Awamleh & Gardner, 1999; Den Hartog, & Verburg, 1997). For instance, DSM's mantra of "Proudly found elsewhere,"[3] Procter & Gamble's "Look to connect and develop before you research and develop"(Cloyd & Euchner, 2012, p. 16), and Henkel's "We borrow with pride"[4] are excellent examples of the skillful use of slogans in sparking change. The importance of storytelling is noted by Amazon's CEO Jeff Bezos, who stated in an interview that "there are stories we tell ourselves internally about persistence and patience, long-term thinking, staying focused on the customer" (Hansen, Ibarra, & Peyer, 2013, p. 85), arguably in order to keep employees on track and unwavering. Finally, artful communicators instill a sense of shared identity (e.g., "We are Googlers"[5]) and do so by, for instance, frequently using collective pronouns such as "we," "us," and "ours" (Stam et al., 2010).

Second, effective vision communicators use an expressive communication style bringing their conviction and intensity of emotions (e.g., optimism and enthusiasm) to the fore via the skillful use of voice (inflection and pauses), gestures, facial mimicry, and body language (Antonakis, Fenley, & Liechti, 2012). The use of emotional displays by leaders has been shown to be highly effective in securing buy-in from subordinates via processes of emotional contagion, whereby subordinates start feeling the emotions displayed by their leaders (Damen, van Knippenberg, & van Knippenberg, 2008). Most importantly, visions that are communicated with genuine passion and intent are more likely to generate internal commitment, which is crucial for perseverance and sustained performance in the face of obstacles (tough schedule, lack of resources, etc.) (Yukl, 2010).

Third, visions that are communicated frequently and are integrated across a variety of channels are more likely to be accepted and internalized (Argenti, Howell, & Beck, 2005; Yukl, 2010). Effective executives understand that they are the face and voice of the organization and that communication cannot be outsourced to the communication department. Frequent, personal, interactive communication, whether during company meetings, team-building exercises, or around the water cooler, tends to be the most effective because it actively engages the listeners and allows for questions and the voicing of concerns (Yukl, 2010). At the same time, making use of an integrated multiplatform approach—via the corporate website, videotaped speeches, internal documentation, e-mails, personal blogs, and social media—can help reinforce and strengthen the message.

Promote Shared Ownership

Successfully getting people on board in the shift from closed to open innovation goes beyond a top-down approach, in which leaders develop a vision of OI behind closed doors and then unleash a perfect storm of communication on a hapless group of employees. Effective leaders understand that large-scale acceptance and adoption of any vision of change requires distributing ownership of that vision as broadly as possible (Ready & Conger, 2008; Yukl, 2010).

There are multiple ways to achieve this aim. First, to enable a sense of shared ownership, key stakeholders whose support is crucial for OI adoption should be involved at the very early stages. Depending on the organization, these could be members of the top management team, senior executives from the different business units, influential union leaders, as well as highly networked employees, regardless of their position in the formal organizational chart. Based on our own conversations with an executive trying to bring OI to her organization, it might even be helpful to include one or two of the most vocal opponents of OI, especially if they are well-respected members of the community. In her words, "The two of them wore us down with reasons why OI wouldn't work, but once they were on board, we knew that we had crossed the Rubicon."

The key advantage of mobilizing the support of such a wide array of stakeholders at the early stages of OI adoption is that those involved in the co-creation process are more likely to have a sense of shared ownership of the change and hence are more likely to be committed to it (Kotter & Schlesinger, 1979). However, there are several additional benefits to coopting such a variety of constituents early on in the process. One advantage is that these constituents can voice different views and perspectives on the issue and can articulate their concerns about potential challenges lying ahead. To this end, leaders should actively encourage an open exchange of opinions and explicitly promote the expression of conflicting points of view (Yukl, 2008). Some companies, such as IBM, Shell, Anheuser-Busch, and 3M, use formalized methods to spark the discussion of conflicting points of view, such as dialectical inquiry methods (Schweiger, Sandberg, & Ragan, 1986) or devil's advocacy (Schwenk, 1984). Research shows that such "constructive controversy" contributes to successful decision-making and innovation (Chen & Tjosvold, 2002; Tjosvold & Yu, 2007). Furthermore, involving a wide array of constituents builds trust in top management and among the various stakeholders, and it serves as a signaling function to the rest of the organization, communicating that openness and collaboration are lived values and not just professed on paper.

In addition, employees in favor of OI should be given a platform to advocate their support throughout the adoption process. As Worley and Lawler (2006) aptly stated, "Think of the corporation as a community of people spread over miles of hills, fields, and forests. To get everyone moving in a new direction, leaders need to be dispersed across the countryside" (p. 22). Those involved in the early stages of OI adoption are the most obvious line of offense in becoming active champions of OI. Another group of self-evident evangelists is those who already believe in the OI paradigm. They are intrinsically motivated to spread the news about the benefits of OI, thereby increasing support from the bottom up.

These individuals exist in every organization; it is just a matter of finding them and empowering them to speak up by, for instance, granting them interviews in corporate news bulletins.

Show People That You Mean It

> By far the most important [factor in OI adoption] is to have serious and visible support from top management. Without this, nothing can or will happen.
> —Tomas Lackner, *Head of Open Innovation and Corporate Scouting at Siemens Corporate Technology*[6]

For the shift from closed to open innovation to be successful, employees need to believe that leaders really mean it and that OI is not yet another fad that shall also pass. In other words, they need to be convinced that leaders are committed to a long-term change. Too many change efforts fail because leaders run out of steam after the initial blitz of communicating the new vision (Gilley, Dixon, & Gilley, 2008). Effective leaders know that change does not happen overnight and that it requires patience, determination, and resilience. Importantly, they understand that it is a continual process of showing commitment and enabling people to enact the change. Hence, leaders need to do at least two things: walk the talk and show support.

Walk the Talk
Typically, visions of change toward OI are filled with buzzwords such as "openness," "collaboration," "trust," and "commitment." These slogans are often mounted on company plaques and plastered all over the company website, employee newsletters, and PowerPoint presentations. However, without leader behavior that matches the words, they become empty platitudes that (in the best case) are ignored and forgotten or (in the worst case) make leadership the butt of employee jokes and kill the change initiative.

Real change only happens when employees know what is expected of them and when they trust their leaders. To make sure that employees know what is expected of them, leaders should role model desired behaviors (Bandura, 1986). Role modeling is helpful because employees watch their leaders (more than they do others) very closely, seeking signals and cues as to which behavior is desirable (Magee & Galinsky, 2008). The impact of role modeling may be particularly strong during periods of organizational change because these are typically characterized by uncertainty and ambiguity (Schein, 2004). Several examples of effective role modeling can be found in the literature. For instance, when ConocoPhillips aimed to increase cross-organizational knowledge sharing by introducing its Networks of Excellence, senior managers not only endorsed the networks but also actively joined the trenches by answering posts and acting as mentors in the discussions, thus inspiring employees to act the same way (Pugh & Prusak, 2013). Thus, the message is that if you want more collaboration, stop running your own fiefdoms within the top management team and start collaborating. If you want more openness, be open to employee ideas and suggestions. If you want more information sharing, exit stealth mode and share more information.

Research has clearly indicated that walking the talk leads to increased trust in the leader (Palanski & Yammarino, 2009) as well as to increases in employees' willingness to accept change (Armenakis, Harris, & Mossholder, 1993). On the other hand, failing to match words and deeds can be viewed by employees as a form of betrayal and will have disastrous repercussions on the leader's long-term reputation and credibility.

Show Support

Top management support has frequently been touted as being crucial in successful OI implementation (Mortara & Minshall, 2014). In fact, in a cross-industry case study of 36 companies, top management support was identified as being the key enabler of OI implementation (Mortara et al., 2009).

Interestingly, in the OI literature, "showing support" typically tends to be equated with "providing adequate resources for OI adoption," and "resources" in turn are usually taken to mean *financial* resources for project funding. Indeed, resources are critical for innovation (Ekvall & Ryhammer, 1999), and project funding is imperative. However, the potential value of other resources, such as time and human resources dedicated specifically to OI, should not be underestimated. Employees often need to balance their day-to-day job with involvement in OI projects for which they frequently have to learn new skills (Nakagaki, Aber, & Fetterhoff, 2012). This leads to conflicting demands on an already limited resource: time. It also implicitly indicates to employees that OI is not a top priority for management. Therefore, to increase the odds of successful OI adoption, it is crucial to explicitly incorporate OI activities into employees' job descriptions and set dedicated time aside for them. In addition, dedicated OI teams, armed with a clear mandate from the top and a budget, can help successfully implement OI by establishing new practices within the firm and by creating a common language around OI (Mortara & Minshall, 2011). Indeed, an increasing number of companies that are relatively successful with OI, such as Procter & Gamble, DSM, Philips, Siemens, and Unilever, have invested in the creation of OI teams (Mortara & Minshall, 2014). Finally, leaders can free up resources by removing organizational barriers (e.g., red tape and administrative hassles) to OI adoption that sap people's time and energy. For instance, if research and development (R&D) engineers spend a large proportion of their time filling in project status reports, leaders can drop the nonessential ones or simplify the process. This will not only free up employees' time schedule but also send a clear signal that leaders view OI as important and have every intention of embedding it in the organization.

As remarked previously, there is more to showing support than providing resources. Leaders can also engage in supportive behaviors by showing that they value employee contributions and care about their well-being (Kottke & Sharafinski, 1988; Rhoades & Eisenberger, 2002). Although often neglected in the OI literature, this type of support is certainly worthy of attention. Leader supportive behaviors have been linked to several OI-relevant outcome variables, such as increased employee trust in leadership, motivation, cooperation, creativity, performance, and identification with the organization (Amabile, Schatzel, Moneta, & Kramer, 2004; Bass, 1990; Yukl, 2008; Zaccaro, Rittman, & Marks, 2001). Moreover, this perspective underscores the importance of also focusing on the role played by lower-level leaders in supporting OI implementation. Support by direct supervisors is particularly important, given that shifting to OI increases uncertainty regarding expected behaviors, or required skills and knowledge, on the part of employees. Because direct supervisors tend to be employees' most prominent focus of attention with regard to dealing with uncertainty, their support is indispensable.

Leader support behaviors can be divided into more instrumental, task-oriented or more relationship-oriented behaviors, although the boundaries between them tend to get blurred (Amabile et al., 2004; Tierney, 2009). The more instrumental behaviors that are perceived to be supportive by employees usually revolve around providing clarity and structure as well as feedback. Especially in the context of OI, it is imperative that direct supervisors provide goal clarity ("What are we trying to achieve?"), clarify terminology and behavioral expectations ("What does it mean and what do you need to do?"), as well as extend various forms of assistance and guidance throughout the process (Amabile et al., 2004). These types of behaviors not only facilitate learning and performance but also increase motivation and involvement (Amabile, 1996; Rietzschel, Slijkhuis, & Van Yperen, 2014; Yukl, 2008).

In terms of the more relationship-oriented behaviors, sharing information openly, delegating control, consulting with employees about decisions that affect them, and providing them with a platform to voice their opinions and concerns tend to have a positive effect on motivation, acceptance of the change, and trust in the leader (Colquitt et al., 2013; Conger, Kanungo, & Menon, 2000). Other interpersonally sensitive behaviors, such as keeping employees informed about stressful issues, disclosing one's own feelings, and showing concern, empathy, respect, and positive regard, have also been shown to increase trust, motivation, and performance (Amabile et al., 2004; Oldham & Cummings, 1996). Because engaging in any innovation-related behavior is potentially risky, it is also important that leaders provide political support by "standing up" for their employees (Tierney & Farmer, 2004) and by serving as an ambassador for their direct subordinates within the larger organization (Ancona & Caldwell, 1992).

Reshape the Meaning of Success

> A success that has outlived its usefulness may, in the end, be more damaging than failure.
> —PETER DRUCKER *(1973, p. 159)*

A third critical aspect of OI adoption that leaders can significantly influence concerns a redefinition of what it means to be successful. People's deeply ingrained implicit assumptions about "how things are done around here"—that is, beliefs about behaviors that are prevalent and necessary to be successful within the organization—need to be unearthed, challenged, and reframed, especially if they run counter to OI principles. To this end, reward and incentive systems need to be changed and aligned with the new types of behaviors that are desired and expected. Here, we outline how leaders can promote behavioral change by taking a two-pronged approach aimed at unearthing and reframing implicit assumptions and aligning rewards.

UNEARTH AND REFRAME IMPLICIT ASSUMPTIONS
In organizations, a large proportion of individuals' behavior is guided by a set of deeply entrenched tacit beliefs about organizationally appropriate conduct. These beliefs have been developed over time, are socially reinforced, and provide the frame through which information is processed, goals are set, and events are interpreted (Detert & Edmondson, 2011; Senge, 1990). In essence, they are the invisible strings that direct individuals' behavior from out of the shadows. Some of these implicit assumptions may pertain to the type of behavior that will help one climb the career ladder, such as "If you want to make it to the top here, you'd better be a rainmaker." Others may refer to the type of behaviors that will most

certainly either not help or actively hurt one's career, such as "Spending all day talking to people trying to build relationships is jolly good fun, but it will not get you anywhere" or "If you make a mistake, hide it, and if possible blame others for it." Yet others might be specifically related to one's role, such as R&D researchers thinking "I need to solve this problem alone. Asking for help or working with others means admitting failure."

These types of implicit beliefs complicate the shift from closed to open innovation. Some of them may promote behaviors that run counter to the behaviors needed for successful OI implementation (e.g., competition instead of collaboration). Furthermore, because they tend to be largely tacit, it is unlikely that their validity gets challenged, making it difficult to change them. Hence, it is critical that leaders uncover some of these deeply ingrained beliefs and actively challenge and change them where necessary (Fiol, Harris, & House, 1999). In essence, leaders need to nudge people's behaviors in a different direction by providing them with alternative "frames" for processing information and pursuing goals.

To change these implicit beliefs, leaders have language at their disposal as one potent tool they can employ. Language provides a powerful way to frame "reality" and shape behaviors by triggering certain associations regarding desirable/appropriate or undesirable/inappropriate behavior (Thaler & Sunstein, 2008). For instance, researchers found that participants spent a larger proportion of windfall money they had received if it was called a "bonus" rather than a "rebate" (Epley, Mak, & Idson, 2006). Similarly, studies have shown that the labeling of a prisoner's dilemma game—a game in which people can choose to either compete or cooperate—significantly influenced participants' behavior. For instance, calling a prisoner's dilemma game a "community game" led approximately 70% of participants to cooperate, whereas calling the same game a "Wall Street game" led only approximately 30% to cooperate (Liberman, Samuels, & Ross, 2004).

Some case studies also support the idea that leaders can effectively make use of language to shape their employees' behavior. For instance, to change a culture riddled by silence and blame, the COO of a children's hospital introduced a policy of "blameless reporting" in which words such as "errors" and "investigations" were replaced by words such as "accidents" and "analysis" (Garvin, Edmondson, & Gino, 2008). Similarly, to promote more creativity among their staff in the Consumer Marketing department, leaders at Facebook changed the name of the department to "Creative Marketing." This had an immediate effect on the group because people now viewed it as part of their job to be creative (Seelig, 2012). Likewise, OI leaders at Roche Diagnostics discovered that one of the main barriers to OI adoption among their research staff was that they tended to label themselves as "problem solvers." In the Roche Diagnostics context, this implied that researchers needed to solve problems on their own and that searching for help meant admitting failure. As such, this label clearly inhibited collaboration and precluded the use of external knowledge. To overcome this, managers encouraged the use of "solution finders" as a more inclusive label that would favor the search and use of knowledge regardless of where it came from (Nakagaki et al., 2012). Finally, leaders at DuPont skillfully articulated the shift to OI by replacing "We can do it all ourselves" with "Together we can achieve what no one of us individually can accomplish,[7] thus communicating that collaborative behaviors will lead to success.

Align the Rewards

Naturally, successfully reframing implicit assumptions will not be enough to redefine what it means to be successful. Both monetary and nonmonetary rewards need to be consistent with the desired behaviors. It is clearly perilous and counterproductive to reward for A (e.g., number of patents filed) while hoping for B (e.g., an increase in solutions that incorporate external knowledge) (Kerr, 1975).

In terms of monetary rewards, if people continue to get primarily rewarded for in-house development (e.g., number of patents filed and publications), outside ideas will be ignored. Financial incentives need to be revised to reduce the focus on internal innovation and to increase the focus on OI by explicitly introducing incentives for solutions coming from OI. These could, for instance, include rewards for patents that are the result of collaboration, the identification of licensing opportunities, or the initiation of collaborative projects (Remneland-Wikhamn & Wikhamn, 2011;

Salter et al., 2014). In addition, promotion systems would also have to be changed to account for the increased need for collaboration. For example, continuing to promote only the "lone stars" that deliver excellent results but do not collaborate would dissuade employees from investing time in building and maintaining internal and external relationships.

With regard to promoting behavioral change, the value of nonmonetary incentives, such as acknowledgment, praise, and recognition, should not be underestimated. In addition to communicating to employees that their involvement and achievements are valued, nonmonetary incentives can serve as powerful signals to the rest of the organization suggesting that OI is desirable. Nonmonetary incentives could, for instance, consist of an "open innovator of the year" award (Salter et al., 2014) given to the one person who excels at OI or a "collaborator of the year" award for the one individual who has gone out of his or her way to share information and bring people together to find a solution to a problem. In this respect, in the 1990s, Texas Instruments instituted a "not invented here, but I did it anyway" award as part of its push to share best practices internally and increase collaboration (Davenport, De Long, & Beers, 1998). Another example is the strategy employed by Tor Myhren, President and Chief Creative Office at Grey Advertising, who instituted a "heroic failure"[8] award. The award is given out on a quarterly basis to employees who take a major, edgy risk. This sends a clear signal throughout the organization that taking risks is encouraged and that failure is acceptable.

(Over) Invest in Capability Building

> If open innovation is not seen as a long-term capability building exercise, then it will fail.
> —TOMAS LACKNER, *Head of Open Innovation and Scouting at Siemens Corporate Technology*[9]

A fourth and critical aspect of OI implementation that leaders can meaningfully steer concerns the investment in employee capability building. Open innovation requires not only the capability to bring outside knowledge in but also the internal capabilities to review and assess external opportunities and to assimilate them into the organization (Mortara et al., 2009). Indeed, a longitudinal study among 1,170 German firms shows that companies with a strong in-house capacity and extensive cross-functional internal collaboration reap higher benefits from openness than do those without these innovation management capabilities (Salge, Bohne, Farchi, & Piening, 2012). In our view, the development of these internal capabilities requires a concerted investment in developing individuals into effective innovators and enabling and leveraging internal connections among individuals.

Develop Individuals

A shift to OI implies that individuals will be faced with new challenges, such as finding promising opportunities on the outside, successfully selling these internally, navigating the "legal niceties" of what can and cannot be shared with outsiders, and being involved in more cross-functional internal collaboration. Some individuals might already have the necessary abilities and skills, whereas others might not. Throwing these individuals into the water assuming that they will learn to swim is not only unrealistic but also irresponsible. To facilitate successful OI adoption, leaders need to identify individual development needs and ensure that they are addressed.

It is only fair to admit that research into the skills, abilities, and attributes that are critical for OI is still in its infancy. However, some attributes seem logical candidates for development for all employees who are involved in OI activities. These attributes include a learning orientation, adaptability, flexibility, communication, cooperation, proactivity, creativity, and an entrepreneurial attitude (Mortara et al., 2009). Moreover, certain types of knowledge, such as having a broad grasp of the different functions within the company, knowing who knows what, and clearly understanding the company's strategic OI direction, also seem to be helpful in facilitating employees' OI-related activities, regardless of their actual position. Leaders can play a vital role in ensuring that concerted employee development programs, consisting of a combination of training, mentoring, coaching, and job-rotation initiatives, are rolled out throughout the firm. For instance, creating "OI Academies" in which employees in all functions are trained in what OI means and how it can be beneficial to them can help distribute knowledge about OI throughout the firm (Mortara et al., 2009). To underscore the more general value of training in overcoming some barriers to OI adoption, a study among 311 firms showed that both training aimed at deepening existing professional skills and training aimed at increasing creativity skills helped buffer the negative effects of the not-invented-here (NIH) syndrome (De Araújo Burcharth, Knudsen, & Søndergaard, 2014).

Some attributes may need to be developed depending on the specific roles that people hold (e.g., idea scouts, R&D researchers, and OI leaders). For instance, some firms, such as Procter & Gamble, invest in training their tech scouts on how to identify potentially valuable opportunities. Other programs, such as Fiat's "researchers with a briefcase" initiative, focus on training R&D people to become effective gatekeepers by clarifying the borders between information that can be openly shared with outsiders and information that cannot be shared. Yet others, such as Unilever's attempt to roll out a training program for OI managers, focus on developing specific leadership skills needed for OI (Mortara & Minshall, 2014). We also draw attention to another type of role—the "assimilator" role—that is in need of developmental attention but has so far been largely ignored. External ideas need to be translated into a shape and form that will make them palatable internally. Therefore, the organization needs individuals who can help assimilate this external knowledge into the firm (Salter et al., 2014). Specifically, these individuals would need to have the capabilities and expertise to repackage external ideas into a form that appeals to internal stakeholders and to clarify how these ideas can add value internally. In our view, the development of this role is critical for successful OI adoption, given that a large proportion of external ideas brought in by the idea scouts never get absorbed internally.

Facilitate Internal Connections

Many internal barriers to OI—such as limited internal collaboration, difficulties in locating information internally (e.g., the needle-in-a-haystack problem), and problems with assimilating external ideas internally (e.g., the NIH syndrome)—can be traced back to either underdeveloped or underutilized internal networks. Leaders can play a key role in overcoming these barriers by establishing and leveraging a myriad of horizontal and vertical connections among employees across functions. This does not mean that everyone should be connected to everyone else. Rather, it implies that leaders need to take a targeted approach at ensuring that individuals with complementary skills and knowledge are connected to each other.

Leaders can employ a variety of initiatives to foster interpersonal connections that facilitate internal collaboration and the cross-pollination of knowledge. For instance, establishing cross-functional teams or task forces, or instituting horizontal job rotations, can have this effect. As a case in point, 3M managers attribute part of their success with OI to 3M's unofficial job rotation policy, whereby technical employees regularly shift between businesses, labs, and countries (Jaruzelski & Holman, 2011). Other initiatives, such as development programs that promote cohort-based learning and peer coaching, help create not only trust-based personal relationships but also an internal network of peers who can act as advisors (Schweer, Assimakopoulos, Cross, & Thomas, 2012). To this end, BP's peer assist program—a process whereby peer groups of managers are used to drive knowledge sharing and learning—is widely known to help collaboration and decrease the NIH syndrome (Ghoshal & Gratton, 2002). Finally, by identifying and co-opting the internal bridge-builders—the people who know where certain pockets of expertise reside within the company and are willing to connect people to each other—leaders can mitigate the inability to locate internal knowledge.

Keeping the context of open innovation in mind, internal connections can also increase the odds of internal absorption of external ideas. In order to achieve this, leaders would benefit from connecting idea scouts to internal idea brokers. For ideas from the outside to be accepted internally, they not only need to be translated in a form that is palatable internally but also need to reach the right persons (Salter et al., 2014; Whelan et al., 2011). Typically, idea scouts do not have the required internal social standing or the distribution network to effectively broker their ideas. In contrast, successful idea brokers are the "go to" individuals within a company—they are at the center of a large social network, know who knows what, and have the social capital to bring external ideas to the right people. By connecting idea scouts to idea brokers, the odds of successful assimilation of external ideas improve dramatically (Whelan et al., 2011).

CONCLUSIONS

In this chapter, we argued that a micro-level perspective, focused on the role of leadership in the internal implementation of OI programs, is indispensable for an adequate understanding of OI and for the successful transition from a closed to an open innovation model. To this end, we presented a number of domains in which leaders can help to overcome some of the most common internal impediments to OI adoption, clustered around employees' motivation and ability to engage in OI activities. Specifically, we stressed the critical role that leaders can play in (1) getting people on board by creating and promoting a shared vision of OI within the organization, (2) earning employee commitment to OI by showing in words and deeds that it is more than just a new flavor of the month, (3) reshaping the definition of success within the organization by challenging assumptions about appropriate behaviors and aligning the reward and incentive systems, and (4) contributing to internal capability building by developing individuals and facilitating the development of internal networks. Moreover, we provided some practical strategies that leaders can employ in each of these domains.

As stated previously, our discussion of the role of leadership in facilitating the successful embedding of OI principles in the organization is not meant to be exhaustive, but we hope to have underscored the importance of acknowledging and accommodating motivational and capability factors in any large-scale change initiative from a closed to an open innovation model. Certainly other factors affect the internal success of OI adoption efforts, such as creating the right systems, structures, and processes and ensuring that the more general organizational culture favors OI (Chesbrough & Bogers, 2014; Mortara & Minshall, 2014). Although different authors and practitioners may stress the importance of one factor over the others, it is critical to think about them as being interlocking parts of a larger system and not as stand-alone issues that can be addressed in isolation. Changing any part of a larger system, such as its structure or its processes, will inevitably affect other areas and might give rise to unanticipated consequences. Effective leaders understand that it is arduous work to achieve any large-scale change and work hard on aligning strategy with systems, processes, and structures—all the while being mindful of the effects that any of these changes will have on the people that are affected by them. They also understand that success will not happen overnight and that there will be some trial and error along the way.

Embedding OI into an organization will not just happen. If it is to become part and parcel of the organization, it cannot be left to serendipity and the grassroots efforts of a few OI champions. Rather, it needs to be nurtured and managed as a deliberate act. It is essential that leaders do not sit back and wait for employees to become innovative; instead, they must actively extend an open invitation to open innovation.

Notes

1. Retrieved from http://www.15inno.com/2011/02/24/oppversusthreats.
2. Retrieved from http://www.research.philips.com/open-innovation.
3. Retrieved from http://www.dsm.com/corporate/about/innovation-at-dsm/open-innovation.html.
4. Retrieved from http://www.henkel.com/com/content_data/141926_2009.09.02_10_Investor_Day_MueKi_Leading_in_Innovation.pdf.
5. Retrieved from http://www.google.com/about/company/facts/culture.
6. Retrieved from http://www.innovationmanagement.se/2013/09/05/open-innovation-an-integrated-tool-in-siemens.
7. Retrieved from http://openinnovation.berkeley.edu/Randolph_Gusch_9_17_12.pdf.
8. Retrieved from http://grey.com/us/culture.
9. Retrieved from http://www.innovationmanagement.se/2013/09/05/open-innovation-an-integrated-tool-in-siemens.

REFERENCES

Amabile, T. M. (1996). *Creativity in context*. Boulder, CO: Westview.

Amabile, T. M., Schatzel, E. A., Moneta, G. B., & Kramer, S. J. (2004). Leader behaviors and the work environment for creativity: Perceived leader support. *Leadership Quarterly, 15*, 5–32.

Ancona, D. G., & Caldwell, D. F. (1992). Bridging the boundary: External activity and performance in organizational teams. *Administrative Science Quarterly, 37*, 634–665.

Antonakis, J., Fenley, M., & Liechti, S. (2012). Learning charisma: Transform yourself into someone people want to follow. *Harvard Business Review, 90*(6), 127–130.

Argenti, P. A., Howell, R. A., & Beck, K. A. (2005). The strategic communication imperative. *MIT Sloan Management Review, 46*(3), 83–89.

Armenakis, A. A., Harris, S. G., & Mossholder, K. W. (1993). Creating readiness for organizational change. *Human Relations, 46*, 681–703.

Awamleh, R., & Gardner, W. L. (1999). Perceptions of leader charisma and effectiveness: The effects of vision content, delivery, and organizational performance. *Leadership Quarterly, 10*, 345–373.

Bandura, A. (1986). *Social foundations of thought and action: A social cognitive theory*. Englewood Cliffs, NJ: Prentice Hall.

Bass, B. M. (1990). *Handbook of leadership: A survey of theory and research*. New York, NY: Free Press.

Boal, K. B., & Hooijberg, R. (2001). Strategic leadership research: Moving on. *Leadership Quarterly, 11*, 515–549.

Chen, G., & Tjosvold, D. (2002). Cooperative goals and constructive controversy for promoting innovation in student groups in China. *Journal of Education for Business, 78*(1), 46–50.

Chesbrough, H. (2003). *Open innovation: The new imperative for creating and profiting from technology*. Boston, MA: Harvard Business School Press.

Chesbrough, H. (2011). *Open services innovation: Rethinking your business to grow and compete in a new era*. San Francisco, CA: Wiley.

Chesbrough, H., & Bogers, M. (2014). Explicating open innovation: Clarifying an emerging paradigm for understanding innovation. In H. Chesbrough, W. Vanhaverbeke, & J. West (Eds.), *New frontiers in open innovation*. New York, NY: Oxford University Press.

Chesbrough, H., & Brunswicker, S. (2013). *Managing open innovation in large firms: Survey report. Executive survey on open innovation 2013*. Retrieved from http://openinnovation.berkeley.edu/managing-open-innovation-survey-report.pdf.

Chesbrough, H., Vanhaverbeke, W., & West, J. (Eds.). (2006). *Open innovation: Researching a new paradigm*. Oxford, England: Oxford University Press.

Chiaroni, D., Chiesa, V., & Frattini, F. (2011). The open innovation journey: How firms dynamically implement the emerging innovation management paradigm. *Technovation, 31*, 34–43.

Christensen, C. M. (1997). *The innovator's dilemma: When new technologies cause great firms to fail*. Boston, MA: Harvard Business School Press.

Cloyd, G., & Euchner, J. (2012, July–August). Building open innovation at P&G: An interview with Gil Cloyd. *Research Technology Management*, 14–19.

Colquitt, J. A., Scott, B. A., Rodell, J. B., Long, D. M., Zapata, C. P., Conlon, D. E., & Wesson, M. J. (2013). Justice at the millennium, a decade later: A meta-analytic test of social exchange and affect-based perspectives. *Journal of Applied Psychology, 98*, 199–236.

Conger, J. A., & Kanungo, R. N. (1998). *Charismatic leadership in organizations*. Thousand Oaks, CA: Sage.

Conger, J. A., Kanungo, R. N., & Menon, S. T. (2000). Charismatic leadership and follower effects. *Journal of Organizational Behavior, 21*, 747–767.

da Mota Pedrosa, A., Välling, M., & Boyd, B. (2013). Knowledge related activities in open innovation: Managers' characteristics and practices. *International Journal of Technology Management, 61*(3–4), 254–273.

Dahlander, L., & Gann, D. M. (2010). How open is innovation? *Research Policy, 39*(6), 699–709.

Damen, F., van Knippenberg, B. M., & van Knippenberg, D. (2008). Affective match in leadership: Leader emotional display, follower positive affect, and follower performance. *Journal of Applied Social Psychology, 38*(4), 868–902.

Davenport, T. H., De Long, D. W., & Beers, M. C. (1998). Successful knowledge management projects. *MIT Sloan Management Review, 39*(2), 43–57.

Davis, J. P., & Eisenhardt, K. M. (2011). Rotating leadership and collaborative innovation: Recombination processes in symbiotic relationships. *Administrative Science Quarterly, 56*(2), 159–201.

De Araújo Burcharth, A. L., Knudsen, M. P., & Søndergaard, H. A. (2014). Neither invented nor shared here: The impact and management of attitudes for the adoption of open innovation practices. *Technovation, 34*, 149–161.

Den Hartog, D. N., & Verburg, R. M. (1997). Charisma and rhetoric: Communication techniques of international business leaders. *Leadership Quarterly, 8*, 355–391.

Detert, J. R., & Edmondson, A. C. (2011). Implicit voice theories: An emerging understanding of self-censorship at work. *Academy of Management Journal, 54*(3), 461–488.

Drucker, P. (1973). *Management: Tasks, responsibilities, practices*. New York, NY: HarperCollins.

Ekvall, G., & Ryhammer, L. (1999). The creative climate: Its determinants and effects at a Swedish University. *Creativity Research Journal, 12*, 303–310.

Elmquist, M., Fredberg, T., & Ollila, S. (2009). Exploring the field of open innovation. *European Journal of Innovation Management, 12*(3), 326–345.

Emrich, C. G., Brower, H. H., Feldman, J. M., & Garland, H. (2001). Images in words: Presidential rhetoric, charisma and greatness. *Administrative Science Quarterly, 46*, 527–557.

Epley, N., Mak, D., & Idson, L. C. (2006). Bonus or rebate? The impact of income framing on spending and saving. *Journal of Behavioral Decision Making, 19*(3), 213–227.

Fiol, C. M., Harris, D., & House, R. (1999). Charismatic leadership: Strategies for effecting social change. *Leadership Quarterly, 10*, 449–482.

Foss, N. J., Husted, K., & Michailova, S. (2010). Governing knowledge sharing in organizations: Levels of analysis, governance mechanisms, and research directions. *Journal of Management Studies, 47*, 455–482.

Garvin, D. A., Edmondson, A. C., & Gino, F. (2008). Is yours a learning organization? *Harvard Business Review, 86*(3), 109–116.

Ghoshal, S., & Gratton, L. (2002). Integrating the enterprise. *MIT Sloan Management Review, 44*(1), 31–38.

Giannopolou, E., Yström, A, & Ollila, S. (2011). Turning open innovation into practice: Open innovation research through the lens of managers. *International Journal of Innovation Management, 15*(3), 505–524.

Gilley, A., Dixon, P., & Gilley, J. W. (2008). Characteristics of leadership effectiveness: Implementing change and driving innovation in organizations. *Human Resource Development Quarterly, 19*(2), 153–169.

Hansen, M. T., Ibarra, H., & Peyer, U. (2013). The best-performing CEOs in the world. *Harvard Business Review, 91*(1), 81–95.

Huizingh, E. K. R. E. (2011). Open innovation: State of the art and future perspectives. *Technovation, 31*, 2–9.

Jaruzelski, R., & Holman, B. (2011, March–April). Casting a wide net: Building the capabilities for open innovation. *Ivey Business Journal*. Retrieved from http://iveybusinessjournal.com/topics/innovation/casting-a-wide-net-building-the-capabilities-for-open-innovation#.U9E5VVYkIds.

Kanter, R. M. (1983). *The change masters*. New York, NY: Simon & Schuster.

Kerr, S. (1975). On the folly of rewarding for A while hoping for B. *Academy of Management Journal, 18*, 769–783.

Kirkpatrick, S. A., & Locke, E. A. (1996). Direct and indirect effects of three core charismatic leadership components on performance and attitudes. *Journal of Applied Psychology, 81*, 36–51.

Kotter, J. P., & Schlesinger, L. A. (1979). Choosing strategies for change. *Harvard Business Review, 57*(2), 106–114.

Kottke, J. L., & Sharafinski, C. E. (1988). Measuring perceived supervisory and organizational support. *Educational and Psychological Measurement, 48*, 1075–1079.

Laursen, K., & Salter, A. J. (2006). Open for innovation: The role of openness in explaining innovation performance among UK manufacturing firms. *Strategic Management Journal, 27*(2), 131–150.

Liberman, V., Samuels, S. M., & Ross, L. (2004). The name of the game: Predictive power of reputations versus situational labels in determining prisoner's dilemma game moves. *Personality and Social Psychology Bulletin, 30*(9), 1175–1185.

Magee, J. C., & Galinsky, A. D. (2008). Social hierarchy: The self-reinforcing nature of power and status. *Academy of Management Annals, 2*, 351–398.

Mortara, L., & Minshall, T. (2011). How do large multinational corporations implement open innovation? *Technovation, 31*, 586–597.

Mortara, L., & Minshall, T. (2014). Patterns of implementation of OI in MNCs. In H. Chesbrough, W. Vanhaverbeke, & J. West (Eds.), *New frontiers in open innovation*. Oxford, England: Oxford University Press.

Mortara, L., Napp, J. J., Slacik, I., & Minshall, T. (2009). *How to implement open innovation: Lessons from studying large multinational companies*. Cambridge, England: University of Cambridge Institute for Manufacturing.

Nakagaki, P., Aber, J., & Fetterhoff, T. (2012, July–August). The challenges in implementing open innovation in a global innovation-driven corporation. *Research Technology Management*, 32–38.

Oldham, G. R., & Cummings, A. (1996). Employee creativity: Personal and contextual factors at work. *Academy of Management Journal, 39*, 607–634.

Palanski, M. E., & Yammarino, F. J. (2009). Integrity and leadership: A multi-level conceptual framework. *Leadership Quarterly, 20*, 405–420.

Pugh, K., & Prusak, L. (2013). Designing effective knowledge networks. *MIT Sloan Management Review, 55*(1), 79–88.

Ready, D. A., & Conger, J. A. (2008). Enabling bold visions. *MIT Sloan Management Review, 49*(2), 70–76.

Remneland-Wikhamn, B., & Wikhamn, W. (2011). Open innovation climate measure: The introduction of a validated scale. *Creativity and Innovation Management, 20*(4), 284–295.

Rhoades, L., & Eisenberger, R. (2002). Perceived organizational support: A review of the literature. *Journal of Applied Psychology, 87*(4), 698–714.

Rietzschel, E. F., Slijkhuis, J. M., & Van Yperen, N. W. (2014). Close monitoring as a contextual stimulator: How need for structure affects the relation between close monitoring and work outcomes. *European Journal of Work and Organizational Psychology, 23*, 394–404.

Rigby, D., & Bilodeau, B. (2013). *Management tools and trends 2013*. Retrieved from http://bain.com/Images/BAIN_BRIEF_Management_Tools_%26_Trends_2013.pdf.

Roberto, M. A., & Levesque, L. C. (2005). The art of making change initiatives stick. *MIT Sloan Management Review, 46*(4), 53–60.

Salge, T. O., Bohne, T. M., Farchi, T., & Piening, E. P. (2012). Harnessing the value of open innovation: The moderating role of innovation management. *International Journal of Innovation Management, 16*(3), 1–26.

Salter, A., Criscuolo, P., & Ter Wal, A. L. J. (2014). Coping with open innovation: Responding to the challenges of external engagement in R&D. *California Management Review, 56*(2), 77–94.

Schein, E. H. (1993). How can organizations learn faster? The challenge of entering the green room. *MIT Sloan Management Review, 34*(2), 85–92.

Schein, E. H. (2004). *Organizational culture and leadership* (3rd ed.). San Francisco, CA: Jossey-Bass.

Schweer, M., Assimakopoulos, D., Cross, R., & Thomas, R. J. (2012). Building a well-networked organization. *MIT Sloan Management Review, 53*(2), 35–42.

Schweiger, D. M., Sandberg, W. R., & Ragan, J. W. (1986). Group approaches for improving strategic decision-making: A comparative analysis of dialectical inquiry, devil's advocacy, and consensus. *Academy of Management Journal, 29*, 51–71.

Schwenk, C. (1984). Cognitive simplification processes in strategic decision-making. *Strategic Management Journal, 5*, 111–128.

Seelig, T. L. (2012). *inGenius: A crash course on creativity.* New York, NY: HarperCollins.

Senge, P. M. (1990). *The fifth discipline: The art and practice of the learning organization.* New York, NY: Doubleday/Currency.

Shamir, B., Arthur, M. B., & House, R. J. (1994). The rhetoric of charismatic leadership: A theoretical extension, a case study, and implications for future research. *Leadership Quarterly, 5,* 25–42.

Slowinski, G., & Sagal, M. W. (2010, September–October). Good practices in open innovation. *Research Technology Management,* 38–45.

Stam, D. A., van Knippenberg, D., & Wisse, B. M. (2010). The role of regulatory fit in visionary leadership. *Journal of Organizational Behavior, 31*(4), 499–518.

Sutton, R. I., & Rao, H. (2014). *Scaling up excellence: Getting to more without settling for less.* New York, NY: Crown Business.

Thaler, R. H., & Sunstein, C. R. (2008). *Nudge: Improving decisions about health, wealth, and happiness.* New Haven, CT: Yale University Press.

Tierney, P. (2009). Leadership and employee creativity. In J. Zhou & C. E. Shalley (Eds.), *Handbook of organizational creativity* (pp. 95–125). New York, NY: Psychology Press.

Tierney, P., & Farmer, S. M. (2004). The Pygmalion process and employee creativity. *Journal of Management, 30,* 413–432.

Tjosvold, D., & Yu, Z. Y. (2007). Group risk-taking: The constructive role of controversy in China. *Group and Organization Management, 32,* 653–674.

van der Vrande, V., Vanhaverbeke, W., & Gassmann, O. (2010). Broadening the scope of open innovation: Past research, current state and future directions. *International Journal of Technology Management, 52*(3–4), 221–235.

Whelan, E., Parise, S., de Valk, J., & Aalbers, R. (2011). Creating employee networks that deliver open innovation. *MIT Sloan Management Review, 53*(1), 37–44.

Worley, C. G., & Lawler, E. E. (2006). Designing organizations that are built to change. *MIT Sloan Management Review, 48*(1), 19–23.

Yukl, G. (2008). How leaders influence organizational effectiveness. *Leadership Quarterly, 19,* 708–722.

Yukl, G. (2010). *Leadership in organizations* (7th ed.). Upper Saddle River, NJ: Prentice-Hall.

Zaccaro, S. J., Rittman, A. L., & Marks, M. A. (2001). Team leadership. *Leadership Quarterly, 12,* 451–484.

8

Innovation Cells

Company Beachheads in Technology Universities

FELIX CARDENAS, TONY DAVILA, AND DANIEL OYON ■

INTRODUCTION TO INNOVATION CELLS

In order to develop disruptive innovation also known as radical innovation, companies often require technology and marketing capabilities beyond their competencies. Consequently, innovative companies organize in new open ways to leverage talent within the firm with external complementary partners (Afuah, 2003; O'Connor & DeMartino, 2006; Schumpeter, 1934). Innovation cells are corporate innovation units that function as military beachheads that secure new technology and marketing insights pursuing promising disruptive technologies and ideas that could turn into commercial blockbusters. If such initiatives prove technologically and commercially feasible, then corporate reinforcements from research and development (R&D) and business units get involved to scale up the new product. These innovation cells operate within leading technology and engineering universities collocated with different actors of the innovation ecosystem, such as researchers, startups, students, venture capitalists, government entrepreneurial agencies, and other large companies.

The innovation cell operates by first discovering new technology and sourcing new ideas by interacting with other actors of the ecosystem. Then companies evaluate the technology feasibility involving the company R&D and innovation cell to assess the development of such new technology. Afterwards, the marketing development begins, in which business units interact with the innovation cell to develop marketing and commercialization plans. Finally, the project gets transferred to a sponsored business unit for commercialization (see Figure 8.1).

The study presented in this chapter is based on more than 40 interviews with different actors of the innovation ecosystem of the École Polytechnique Fédérale de Lausanne (EPFL), a leading Swiss engineering

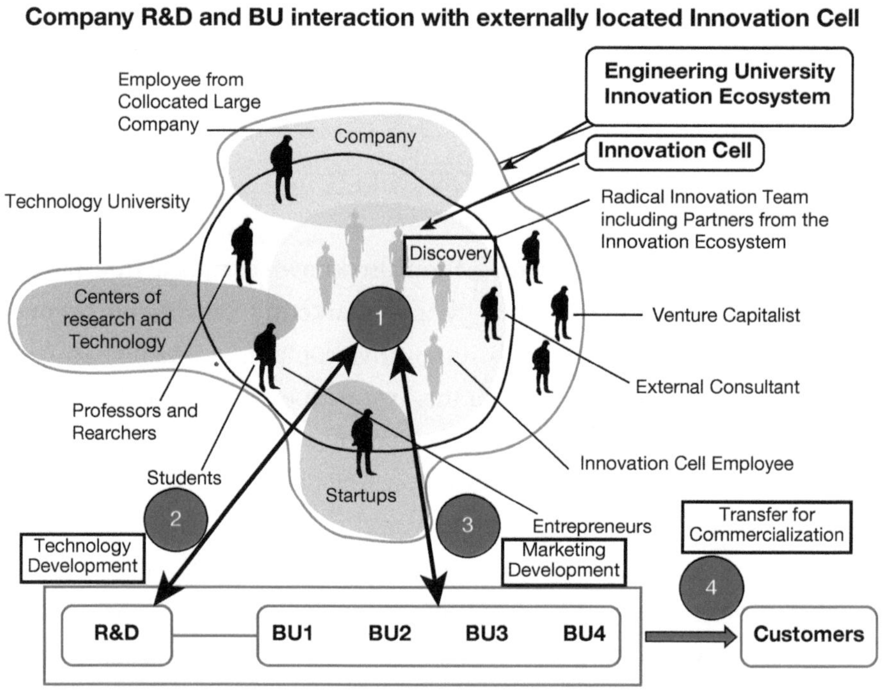

Company R&D and BU interaction with externally located Innovation Cell

1. Discovery: Innovation Cell interacts with researchers, students, startups, venture capitalists, consultants, entrepreneurs, and other large companies to source and develop ideas
2. Technology Development: Company R&D interacts with Innovation Cell to develop technology
3. Marketing Development: Company Business Units interact with Innovation Cell to develop marketing for commercialization
4. Project Transfer to BU for Commercialization. Business Units commercialize new products and services

Figure 8.1 Innovation cell interacting with other actors in a university ecosystem.

university also known as the Swiss Federal Institute of Technology. Six organizations agreed to participate in this study: Credit Suisse, Nestlé, Cisco, Alcan-Constellium, PSA Peugeot Citroën, and EPFL. This chapter addresses the fundamental aspects of large firm–technology university interactions; the motivation to establish an innovation cell; the operation, organization, and culture of an innovation cell; and evidence of disruptive innovation. The study identifies how large companies recruit key personnel on campus; establish complementary partnerships; and involve themselves in the innovation ecosystem to identify, pursue, develop, and transfer disruptive technologies into their organizations.

The innovation cell needs to have certain competencies enabling it to be capable of discovering new technology opportunities inside the university. Moreover, it must be able to work with its corporate R&D group and business units to develop the technology and reach a proven concept prototype to validate viability to be manufactured in large scale. Should the corporation and university not be able to develop the technology, then the innovation cell will need to develop partnerships in which a clear definition and rights of the intellectual property need to be specified to avoid future tensions among the partners regarding intellectual property ownership (Ahuja & Lampert, 2001; Avnimelech & Teubal, 2006; Dushnitsky & Lenox, 2006; Guth & Ginsberg, 1990; Leamon & Lerner, 2012). See Figure 8.2 for innovation cell competencies. It is important to note that innovation cells face two main challenges: creating a stimulating environment to explore new technologies and transferring these technologies back to the company. For the latter, it is very important to have a sponsor business unit early in the development stage (Chesbrough, 2002a).

In order to establish an innovation cell, companies need to create an independent unit apart from the main organization that tends to have a short-term mentality, affecting the long-term horizon required to develop disruptive innovation. Not only does the innovation cell need to be located inside a campus of a leading technology engineering university but also the chosen university has to strive for industrial liaison collaborations and technology transfer partnerships to enable the development of market disruptive innovations. The innovation cell works as

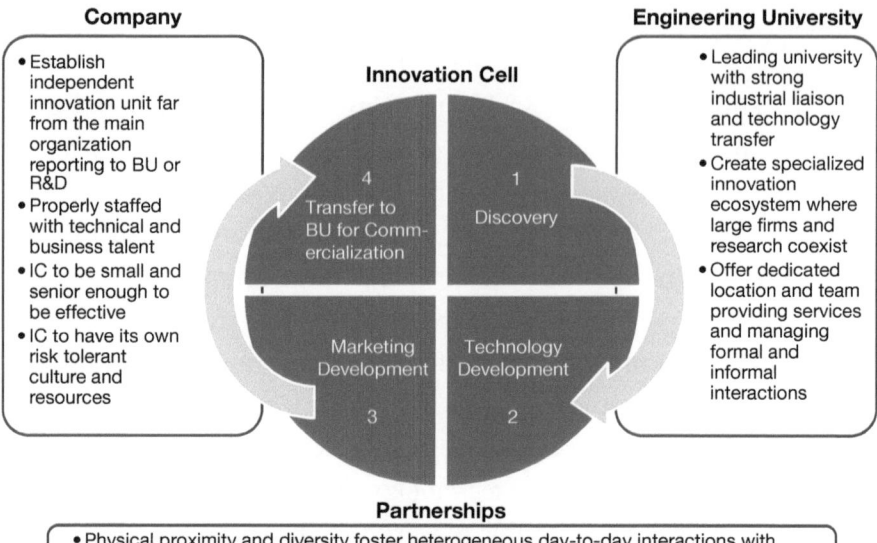

Figure 8.2 Innovation cell competencies.

a membrane bridging the university scientific environment and the corporation business units. The innovation cell needs to be properly staffed with engineering and business-savvy personnel. The team must have senior members to communicate effectively and get access to corporate resources. The innovation cell team will have its own culture and performance criteria. Innovation cells usually report to headquarters or senior managers at business units; seldom do they report to R&D because R&D tends to be risk averse and lacks commercial capabilities and marketing experience. A roadmap to establish an innovation cell is presented in Figure 8.3.

INNOVATION CELLS, R&D, BUSINESS UNITS, AND OPEN ORGANIZATIONAL STRUCTURES

Companies are exploring new ways to be first at identifying new technologies, benefiting from quickly bringing these to market and enjoying a first

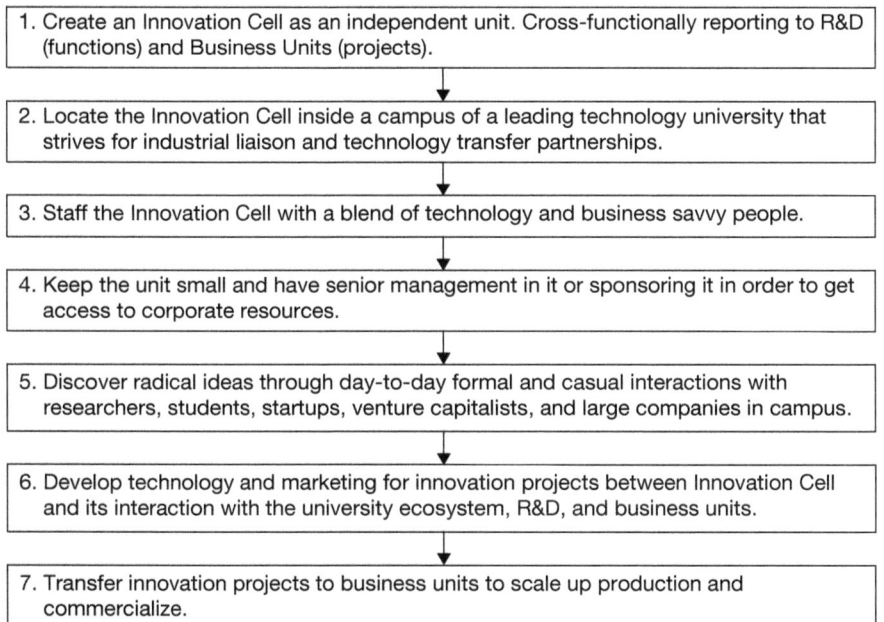

Figure 8.3 Roadmap: How to introduce innovation cells.

mover advantage. In some cases, if they are not the first ones to exploit a nascent technology, then they learn from the mistakes of other early entrants in the market and accelerate the deployment of new versions of improved products (Markides & Geroski, 2005). The challenge for these large companies is how to hunt, capture, and deploy these disruptive ideas to market. R&D traditionally has been the main internal approach to generate innovation. However, open innovation[1] has proven to be an effective additional element of innovation management. Companies have been experimenting with open organizational structures to manage their disruptive innovation projects. For instance, Fiat, the Italian car company that owns Chrysler, claims that this type of open organizational structure has helped it reduce the cost of its R&D activities while maintaining future growth opportunities (Chesbrough & Rosenbloom, 2002). Fiat's open organizational structure allowed it to establish collaborations among its 850 research professionals employed at its Centro Richerche Fiat and more than 150 universities and 1000 industrial partners worldwide. Another example is Procter & Gamble, which changed its innovation approach from traditional R&D to a "Connect & Develop" program to benefit from the ideas of millions of external people throughout the world, increasing its R&D productivity by more than 60% and accelerating its innovation process (Chesbrough & Tachau, 2002).

In contrast to incremental innovation that depicts small growth opportunities and low market and technology risks, disruptive innovation deals with a significant amount of uncertainty exploring ideas that have the potential to disrupt existing markets. If successful, it provides companies with sustainable competitive advantage and high economic returns (Day, 2007; Ettlie, 1983; Markides & Geroski, 2005). Most companies' innovation portfolios have approximately 90% of their products developing incremental innovation merely aspiring to maintain their position in the market (Cooper, 2003). However, these efforts rarely generate the growth platforms that revitalize companies' growth path. Disruptive innovation, on the other hand, accounts for approximately 10% of innovation portfolios, but it accrues more than 60% of all profits from new product development (Kim & Mauborgne, 1999).

Companies have benefitted from collaborative organizational open structures enabling them to generate disruptive innovation (Lerner, Schoar, & Wongsunwai, 2007). These collaborative organizational open structures leverage both internal and external technological resources (Rozanov, 2005). An alternative to achieve this open organizational structure is to locate a heterogeneous team in the vicinity of a diverse multidisciplinary innovation ecosystem with a vibrant startup community (Griliches, 1979). Furthermore, this organizational approach (Dushnitsky & Lenox, 2005; Hall, 1992; Himmelberg & Petersen, 1994) tapping into novel ideas is more efficient when dedicated innovation units are physically separated from the mainstream corporate organization and business units that often pursue immediate short-term results instead of a long-term vision required to develop disruptive innovation (Chesbrough, 2002b).

R&D and innovation cells differ mainly because R&D is mostly conducting fundamental research, whereas innovation cells are technologically capable but also have a strong component of marketing and business development.

Moreover, R&D aims at developing long-term research, whereas innovation cells pursue disruptive innovation in the short and medium term.

In order to determine whether the organizational structure to develop a specific innovation project requires an innovation cell, companies have to consider two characteristics of the project: the stage of product development and the level of "disruptiveness" of such product or service. The level of disruptiveness is determined by how close a specific innovation project is to existent products and processes or how far it is from current capabilities and if it could create new growth platforms with new product categories. The level of disruptiveness will determine whether the project remains within R&D and a business unit or if it requires support from an innovation cell (Figure 8.4). The horizontal axis in Figure 8.4 refers to the stage of product development, ranging from an early stage, "exploratory," to a later stage, "industrialization and market launch." Exploratory means that R&D is exploring at the laboratory scale and the technology has not yet been demonstrated, whereas industrialization and market launch is a developed stage of industrialization. For instance, prototypes

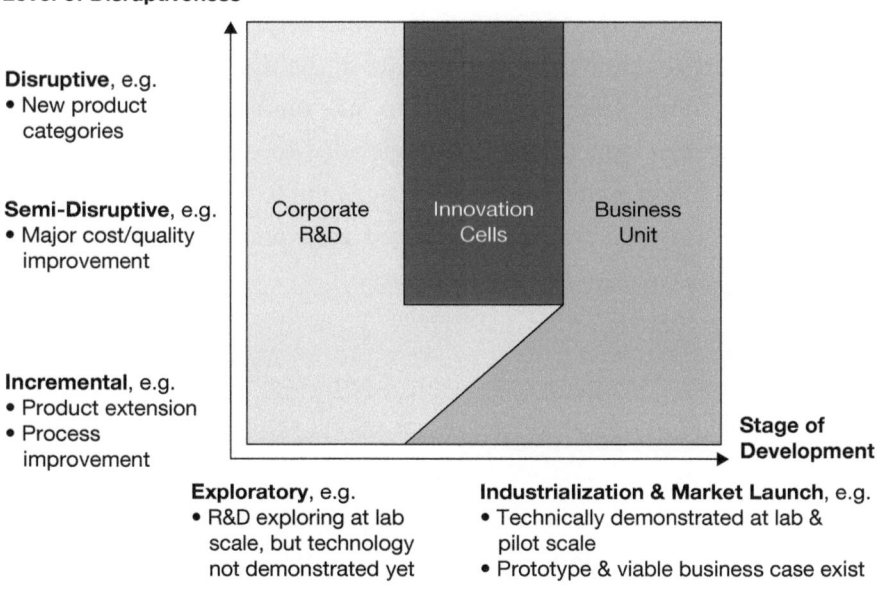

Figure 8.4 Organization contingent to "disruptiveness" and development stage.

are available and there is evidence that there is a viable commercial business case with real demand for the product. The vertical axis in Figure 8.4 indicates the level of disruptiveness, which has at its origin incremental projects such as a product extension or minor improvements. Next, semi-disruptive innovations can be projects of major cost reduction or quality improvement. Finally, a high level of disruptiveness includes innovation that reflects a new product category with significant market and technology risks as well as high growth potential. R&D and business units are able to develop and launch incremental innovative products without the involvement of innovation cells. However, for projects with a higher level of disruptiveness, in which marketing and technology uncertainty prevail, innovation cells participate more actively due to their technology and business development capabilities. It is important to note that when developing disruptive innovation, companies need to find a recipient business unit willing to sponsor the innovation project or to commit to form a new business unit (Markides & Oyon, 2000). For example, for Nestlé, Nespresso represented a disruptive innovation to make coffee out of an aluminum foil capsule cartridge, a technology that was based on the work of a group of external engineers. Nestlé brought into the firm the external coffee cartridge invention, which required hardware coffee machine expertise that Nestlé did not have at the time. Because no existing business unit at Nestlé was willing to take on Nespresso's disruptive idea (Markides & Oyon, 2000), Nestlé decided to establish a new business unit with full support from top management to tackle organizational resistance because Nespresso was viewed as a potential cannibalizing agent to Nestlé's traditional coffee products.

ENGINEERING TECHNOLOGY UNIVERSITIES AND EPFL'S INNOVATION SQUARE

Studies highlight the benefits of physical collocation and working in multidisciplinary networks to facilitate the interactions among complementary actors that can generate disruptive innovation (Duchesneau, Cohn, & Dutton, 1979). This multiorganizational ecosystem is similar to that of an ocean coral reef, where startups are like small fish that receive nourishment from marine plants and protect themselves from threats to survive and grow. This concept has been documented at universities such at the University of Texas (UT) at Austin, where university-based accelerators bring together startups, experienced business leaders, researchers, government officials, large companies, and venture capitalists. This ecosystem provides startups with technical expertise, commercial business development, and access to capital (Markman, 2012).

Novel technologies are often found in technology engineering universities; companies aware of this pool of knowledge are exploring ways to tap into this university talent (Cohen, Goto, Nagata, Nelson, & Walsh, 2002; Munson & Pelz, 1979). In the United States, leading engineering and technology universities, such as the the Massachusetts Institute of Technology (MIT), Stanford, UT Austin, and California Institute of Technology (Caltech), have a reputation of collaborating with industrial partners to develop products and services. In Europe, universities such as the KTH Royal Institute of Technology in Sweden, Oxford and Cambridge Universities and Imperial College in the United Kingdom, ETH and EPFL in Switzerland, and the Technical University of Munich in Germany have established partnerships with companies in areas such as information technology, aerospace, material science, bioscience, and electrical, chemical, and mechanical engineering. This is also applicable in emerging markets. For instance, Embraer, a Brazilian aeronautics corporation that is one of the world's leading aircraft manufacturers, has benefited from its long-term partnerships with research centers and engineering universities such as Instituto Tecnológico de Aeronáutica and Departamento de Ciência e Tecnologia Aeroespacial.

Switzerland has an excellent reputation for research and technology transfer from research centers to companies. It consistently ranks as one of the top countries for innovation[2] (Sala-i-Martín et al., 2012). It has developed strong academic engineering programs, such as ETH in Zurich and EPFL in Lausanne.[3] The Swiss government supports companies embarking on innovative initiatives through dedicated government organizations such as the Commission for Technology and Innovation (CTI) that bestows support to R&D projects and entrepreneurs. CTI encourages the transfer of knowledge and technology between higher-education institutions and companies pursuing innovation.

The EPFL has five schools: computer and communication sciences; life sciences; architecture and civil environmental engineering; basic sciences; and engineering, which includes materials, electronics, electrical, mechanical, and micro engineering. Some examples of innovative initiatives in which the EPFL has been involved are the Alinghi sailboat that won the America's Cup and the Solar Impulse project housed at EPFL to operate a purely solar-powered airplane capable of flying around the world.

In 2008, the EPFL established Innovation Square, a facility within the campus designed to house large companies, startups, and small and medium innovative enterprises. The goal has been to make these companies interact with EPFL technological capabilities to generate disruptive innovation. EPFL believes that an open organizational structure will enable companies to learn about nascent technologies and benefit from the diverse perspectives that different partners bring to the table. EPFL has been able to attract large companies due to its high-quality research, attractive intellectual property policy, recruitment of international highqualified engineers, and a technologically advanced laboratories infrastructure. In addition to Innovation Square, the Rolex Learning Center is an important element of the ecosystem. The Rolex Learning Center on campus represents a laboratory for learning. It functions as a business–scientific hub open to both students and the public. It provides services ranging from libraries to social study spaces and restaurants. The Rolex Learning Center is a place where traditional boundaries between disciplines are broken down, where multidisciplinary heterogeneous thinkers and doers meet to envision new technologies exchanging ideas to generate innovation. A conference center was also built on campus to facilitate the interaction with the innovation ecosystem actors and the general public. It showcases the leading technologies developed at the EPFL. Due to the campus layout and proximity among the ecosystem actors, there is an active exchange of ideas and initiatives to work together in formal and informal ways. A layout including Innovation Square, the conference center, and the Rolex Learning Center is presented in Figures 8.5 and 8.6.

EPFL is formed by five schools: Computer and Communication Sciences, Life Sciences, Architecture and Civil Environmental Engineering, Basic Sciences, and Engineering which includes Materials, Electronics, Electrical Engineering, Mechanical and Micro Engineering.

Figure 8.5 Technological capabilities and modern business development facilities at EPFL.

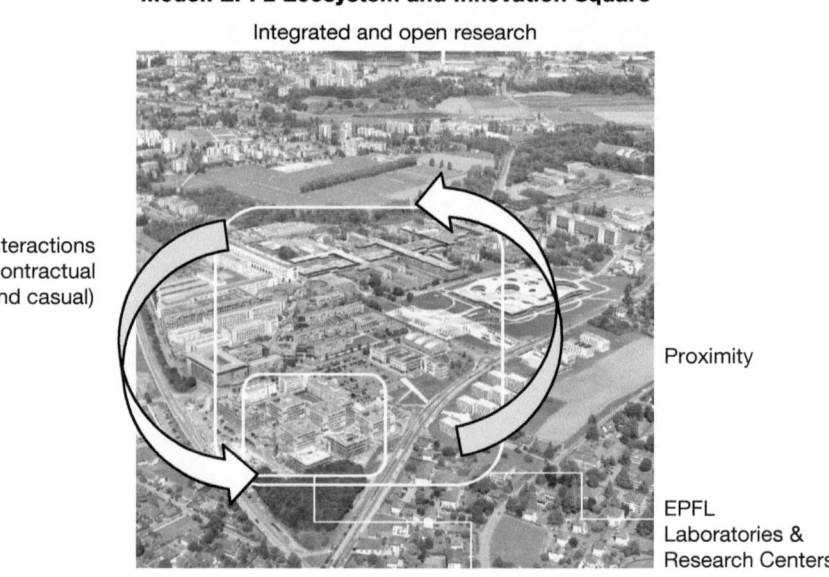

Figure 8.6 EPFL ecosystem enabling formal and informal interactions due to proximity.

Innovation Square is a place where scientists and industrialists can build relationships and feed their creativity. Geographic proximity is a stimulus for cooperation.

—Vice President for Innovation and Technology Transfer at the EPFL

FORMAL AND INFORMAL INTERACTIONS

Previous research indicates that innovation development requires a heterogeneous team exposed to a diverse environment with frequent formal and informal face-to-face interactions with external partners (Davila, Epstein, & Shelton, 2006). We studied how the Innovation Square ecosystem works, with large firms interact in formal and informal ways with other actors at the university campus. Formal interactions operate under a contractual framework in which the EPFL Innovation Square provides research contracts and service contracts linking companies to university laboratories. A research contract comprises intellectual property licensing opportunities, collaboration agreements, and technology transfer from university to companies. These contracts form the formal basis for creating new knowledge in collaboration with companies. Companies use the intellectual property to develop new products, and EPFL keeps the right to publish the findings in scientific journals. Thus, the research results are not kept confidential, which is a delicate matter for a company that wants to protect new discoveries. However, partner companies can apply for a patent to commercially protect the intellectual property before it is divulged publicly. If a patent is granted, EPFL keeps the right to use such patent and offer sublicenses, as long as they are outside the fields of use stipulated in the contract. A service contract is used to provide a service to a company using existing knowledge or laboratory infrastructure to conduct laboratory analyses and simulations.

The Innovation Square ecosystem also offers opportunities for informal interactions. Networking reunions and lunches are an important part of the ecosystem. Reunions are tailored for different groups, such as general managers, marketing and business development managers, and technology experts. At these gatherings, topics relevant to specific groups are discussed. Social events include Innovation & Science on Board, Manager's Lunch, Fun-Rally, Christmas Drink, Get to Know Your Neighbors, and Pizzas & Startups. The objective of Innovation & Science on Board is to create a network to foster exchanges of experiences among executives from different industries. The audience includes managers in charge of innovation cells, CTOs, vice presidents, and directors of innovation. Innovation & Science has 105 members and meets twice a year. Topics include how to manage disruptive innovation in large corporations, how to measure innovation performance, and innovation in emerging countries. Pizzas & Startups are casual gatherings at which startups pitch their innovations and business models to exchange ideas and discuss project collaborations (Figure 8.7).

Interactions: Contractual and Casual

Contractual		Casual
Research contract	**Service contract**	**Social events**
Company pays for a specific research project	Company pays for a specific scientific service (tests, routine analysis, measurements on special equipments, expertise, etc.)	Events and social gatherings*
• Company priority to protect interesting results		• Innovation on Board: Creates a network and fosters exchange of experiences among executives from different industries. Topics presented: How to manage radical innovation in large corporations, how to measure innovation performance.
• IP ownership transferred to Company if patent filed by Company	• All rights on results are transferred to the Company	
• No royalties due later by Company if exploitation	• EPFL remains owner of its methods and tools used for the execution of the service	• Science on Board: Sharing EPFL expertise on specific topics. The audience varies by function in the company. Topics presented: Math and marketing, predict consumers' choices, boosting innovation through Web2.0 & Crowdsourcing
• Free license from Company to EPFL on patented results outside of a defined Field of Use, with the right to grant sublicenses	• Scientific publications and use of results only with prior approval of Company	
• Right for EPFL to publish its scientific results		• Social meetings include: Managers lunch, Fun-rally, x-mas drink, get to know your Neighbors, BarCamp, BBQs, Pizzas & Startups

*Events are targeted and tailored to different individuals within the Innovation Cell function (e.g. general manager, marketing, human resources, engineering, administration)

Figure 8.7 Interactions: formal contractual and informal casual.

Formal and informal interactions help to form collaborations with different members of the ecosystem. Being here leaves no other option but to interact all the time, from bumping in the parking lot to casually discuss an idea, to reviewing project specifics with PhD students in the classroom and laboratories.

—Director, *Cisco Systems International*

LARGE FIRMS ESTABLISHING INNOVATION CELLS AT EPFL

Beginning in 2008, approximately 400 employees of 12 large companies and more than 100 startups with more than 700 employees were located at EPFL's Innovation Square. Companies such as Nokia, Nitto Denko, Logitech, Credit Suisse, Nestlé, Cisco, Medtronic, Debiopharm, Alcan-Constellium, Merck

Serono, and PSA Peugeot Citroën established an innovation cell. Their common goals were to leverage their internal R&D through external collaborations for joint research initiatives, discover new business opportunities, recruit new talent, organize focus groups, and brainstorm with students. The following sections discuss five companies that agreed to share their experience when establishing an innovation cell at EPFL's Innovation Square.

Alcan Engineered Products (Constellium)

In 2008, Alcan Engineered Products, a leading business-to-business aluminum company now called Constellium after a 2011 divestiture, decided to pursue a strategy to differentiate itself from commodity-alike competitors and transform itself from a "me-too" player to a disruptive innovator. Achieving this transformation required Constellium to organize itself differently. R&D had traditionally delivered incremental innovations related mainly to aluminum alloy development and metal surface finish. In order to embark into disruptive innovation, the company required new relationships with research institutes, consultants, and companies from unrelated business sectors. Constellium searched for a location that could provide it with disruptive ideas. The company chose to be based on EPFL's campus to benefit from the innovation ecosystem. Constellium and EPFL established a multiyear cooperation agreement to identify new disruptive ideas for the development of new lightweight materials.

> EPFL is dynamic and has a reputation for excellence to which we can attest. Interactions with professors and labs take place in an atmosphere of open collaboration. This environment encourages us to be creative and more risk-taking in our innovations.
> —President of Global Aerospace,
> *Transportation & Industry, Constellium*

Constellium established its innovation cell separate from the rest of the organization to avoid a short-term vision and tensions with the rest of the organization. By being at Innovation Square, it benefits not only from formal contractual relationships accessing researchers and laboratory facilities but also from the informal access to other ecosystem actors made possible due to proximity.[4] Although the innovation cell is separated from the rest of the organization, it is not isolated because it needs to transfer technology and commercial knowledge to the business units. The technology transfer occurs once product design specification and customer benefits are verified by the innovation cell. Design review includes demonstration of technical feasibility, and both R&D and the recipient business unit are engaged in the product development process to attest that technical capabilities match market expectations. The project ownership is later transferred to the business unit responsible for solving technical-manufacturing challenges and project implementation timeline from production to commercialization.

By being at Innovation Square, Constellium also experiences a process of "culturization" through its exposure to exploratory project initiatives that not necessarily lead in the short term to a commercially successful product. Constellium Innovation Cells bring together a multidisciplinary team of people from different backgrounds, such as commercial, business development, and R&D, working together to deliver market-ready prototypes and innovative business plans in short time frames. In coordination with business units, innovation cells participate as part of the team to develop the commercial implementation plan, scale up, and launch preparation. It goes as far as identifying the marketing and sales human resources gaps that the business unit has and proposes potential candidates for the commercialization functions. The innovation cell is also responsible for identifying and testing the primary target costumer group along with specifying early adopters and heavy users.

> Alcan Innovation Cells can help business units execute disruptive projects by defining risk management plans, and conducting formal design reviews including demonstration of technical feasibility and commercial-financial viability.
> —Managing Director, *Constellium Innovation Cells*

Constellium learned through past experiences that a mix of centralized and decentralized R&D organization is necessary to balance the resource allocation between short- and long-term projects. In the early 1990s, R&D employees worked at a central technology organization located at the company's headquarters. By the mid-1990s, the company decentralized its technology competence and placed most of its engineers and scientists at the business units. Nevertheless, long-term initiatives started to be neglected because business unit managers were more concerned about meeting quarterly goals than about projects that require 10- to 15-year commitments such as those in the aerospace industry. Consequently, Constellium kept some R&D personnel at the centralized unit to avoid neglecting projects that require such a long-term commitment.

Constellium has several examples of disruptive innovation developed at EPFL. These usually involve external partners. The "Aluminum Foam" is an innovative high-efficient heat exchanger offering superior heat transfer capacity with no joining components while reducing volume and weight. It is composed of salt, dough, water, and aluminum. Together with EPFL's mechanical metallurgy laboratory, Constellium scaled up the first prototype from laboratory samples to a successful factory full-scale production run. This project required the participation of "unfamiliar" external partners from the world of food processing. For instance, innovation cells entered into a collaboration with a world-leading pasta extruder to produce salt dough molds. In this project, innovation cells designed the industrial process, identified partners, identified applications, and assessed market potential. Another example is the partnership between Constellium and 3M to develop advanced surface antidust high-reflecting finishing for lightweight solar concentration energy applications. Another disruptive innovation developed by innovation cells is the "Smart Material" project with the vision to revolutionize the aerospace industry by providing materials that can be used for airplane wings allowing these to adapt their shape to changing flight conditions due to an electromechanical interaction at the material crystalline state. These materials can also "heal" themselves after they crack, due to integrated capsules that can sense their environment and adapt their mechanical properties accordingly. These are all examples of the disruptive innovation projects developed by Constellium Innovation Cells at EPFL with external partners, R&D, and business units.

> By being located at EPFL we live in open innovation. Our connection to EPFL helps us stay informed of the latest scientific discoveries. And by working alongside other leading technology companies we've been exposed to many opportunities to explore disruptive projects through external collaborations.
>
> —Project Manager, *Constellium Innovation Cells*

Peugeot Citroën Group

In 2010, PSA Peugeot Citroën launched an initiative spearheaded by their scientific director to reinforce the link between the company and external open innovation applied research. The initiative was known as Open Lab Resources, which aimed to generate new knowledge to provide the firm with key differentiators appealing to consumers. PSA Peugeot Citroën is a French company with most of its innovation and manufacturing activities in Europe; hence, it decided to establish its Open Lab in Europe. After meeting with several universities, the company found at the EPFL a location that was conducive to innovation with strong technology and engineering capabilities. Moreover, EPFL houses the Transportation Center (TraCE). TraCE is an interdisciplinary center with the mission to be the interface between EPFL and the world for all topics related to transportation and mobility of people and goods. TraCE brings together more than 40 laboratories with a combined workforce of more than 560 researchers. It provides a wide range of competences and expertise to address the increasingly complex challenges associated with transportation.

For PSA Peugeot Citroën, the most valuable assets of being at EPFL are the 320 labs with which it can establish a formal relationship and conduct disruptive innovation projects. It also appreciates that EPFL Innovation Square houses startups that are a resource for fresh, new ideas. Another important factor is the interaction with other large companies; for example, a materials company such as Constellium offers insights to cutting-edge applicable trends in material science and market applications.

The staff of PSA Peugeot Citroën Innovation Cell has business and engineering backgrounds. When asked about the culture and challenges faced when transferring technology from the Innovation Cell to the rest of the organization, the staff explained that PSA Peugeot Citroën business units have a very busy agenda, annual approved budgets, and a rather rigid systematic culture, making it difficult to bring new ideas or spontaneous opportunistic discoveries to the organization's immediate attention. Because car model designs are established years in advance and tend to be rigid, the company's innovation cell searches for a sponsor business unit early during the development of the innovation project that will be responsible for developing the new automotive component in future new models.

Being located on the campus offers PSA Peugeot Citroën a unique opportunity to be exposed to a culture of discoveries catalyzed by formal and informal interactions and collaborations with different members of the ecosystem. Regarding the informal interactions, restaurants and bars located on the campus provide for informal discussions. The company also has several formal collaborations with EPFL, one of which is a 4-year long-term research contract in which PSA Peugeot Citroën pays for a PhD student to develop a doctoral study along with a leading professor. The agreement includes licensing opportunities and technology transfer from the results of the doctoral dissertation. EPFL has the right to publish the findings in scientific journals, and Peugeot Citroën is allowed to apply for a patent in order to commercially protect the intellectual property. Peugeot Citroën and EPFL also have a 1-year short-term industrial grant to conduct technology transfer. The purpose of this financial grant is to execute short-term research projects, and in return Peugeot Citroën has the right to evaluate the results of the project and is given the option to negotiate a fee-based license on the results.

An example of a specific cooperation project between PSA Peugeot Citroën and EPFL involves EPFL's TraCE, Signal Processing Laboratory LTS5, and a related startup called nViso, who's technology involves analyzing a driver's expressions and muscle movements and determining whether the driver is too distracted, too tired, or even too angry to safely control his or her driving. This is particularly relevant because there has been an increasing awareness of the dangers of driving while fatigued. In the United States, 1 in 7 licensed drivers aged 16–24 years admitted to having nodded off at least once while driving in the past year compared to 1 in 10 drivers who confessed to falling asleep during the same period. Safety studies estimate that 15–33% of fatal accidents in the United States are caused by drowsy drivers. The technology uses a camera to capture facial expressions, and it uses software to search for signs of distraction as well as emotions that could indicate that the driver is not capable of driving. It is like a copilot who can not only read one's mood but also take action before a driver's mental condition can put him or her at risk by impairing driving. The aim is to optimally utilize computer vision technologies to improve safety and comfort in cars through more natural human–vehicle interface. Although facial recognition technology has become common for surveillance, applying it in cars presents a unique set of challenges, beginning with where to place the camera so that it does not obstruct the driver's view. Another issue is adapting it to changing lighting conditions, such as when a car goes into a tunnel or at night when active safety systems are most beneficial. The next step is to test the facial recognition technology in real-world conditions. It is currently fitted to a prototype vehicle, and EPFL is refining the technology by increasing the number of images that can be processed. In addition, PSA Peugeot Citroën and EPFL have been working on two other initiatives related to the reduction of CO_2 emissions. One is an ultra-fast charger that enables electric vehicles to fully charge batteries in 5 minutes, and the other is related to the optimization of energy use from the source of fossil fuel to the actual driving experience ("well-to-wheel"), which is being performed through a software simulator at a EPFL laboratory.

Cisco Systems

Cisco is an international leader in networking technology. Cisco's innovation initiatives for products and services are partly responsible for its consistent gross margin greater than 60%. Cisco Development Organization is the unit within Cisco in charge of innovation in its core networking technology as well as video, microprocessors, and virtualization software.

Cisco established an innovation cell at EPFL in 2010 to go beyond existing incremental innovation programs and pursue disruptive innovation to generate new product platforms. Cisco established collaborations with the government entrepreneurial agency CTI to have EPFL join a research initiative to develop new video applications.

> For Cisco to be in the EPFL is a new approach to innovate. This is the first time we set up an office inside a university. This kind of presence is a key example of closer collaboration between science and industry.
> —Innovation Manager, *Cisco*

Despite the 2008–2010 global economic downturn, by 2010 Cisco occupied 800 m² at the EPFL Innovation Square. According to Cisco, EPFL is a place where scientists and business managers can build relationships and feed their culture for creativity. Cisco highlights the casual interactions between members of the teams at informal gatherings such as at the bar Satellite or the Rolex Learning Center located inside the campus. In the event that these informal interactions develop into a more formal initiative, then Cisco puts in place a nondisclosure agreement to protect intellectual property among the parties involved. Cisco also profits from relationships with other tenants at the EPFL Innovation Square. For example, Cisco and Credit Suisse share best practices and methodologies for software testing and development.

Cisco admits that in order to have an effective innovation operation, it must staff its innovation cell with highly qualified talented people available through EPFL's recruitment offices. Since 2010, Cisco has hired more than 20 PhDs from EPFL. Most of them remain in Switzerland, but others have been relocated to other Cisco business units in California and Israel.

The challenge Cisco faces when transferring technology to the rest of its organization is that it has its own R&D, and sometimes new ideas coming from the exterior can received a low level of priority or be subject to "pushback" or a "not invented here" syndrome. Consequently, Cisco makes sure to include early on the development the engineering teams of the business units that will be the recipient of the innovation. As a practice, Cisco's innovation cell chooses to involve colleagues from R&D and sales, such as members of its office in New York City, who are responsible for commercial software for the banking industry.

Cisco has both research and service contracts with EPFL. Cisco prefers not to have long-term projects but, rather, short-term ones that can provide deliverables in a period not to exceed 9 months. Cisco values the top quality of EPFL's professors, as well as existing patents that EPFL can offer as licenses. Cisco is particularly interested in EPFL's capabilities to develop Internet and video applications, which can be accelerated by partnering with EPFL faculty; specifically, they have been working together on fiber-to-home and telepresence initiatives.

> This collaborative technological approach will accelerate and advance the development of products and solutions that are important to Cisco.
>
> —Innovation Manager, *Cisco*

Nestlé

Nestlé's world headquarters are based in Vevey, Switzerland, approximately 30 miles from EPFL. Both Nestlé and EPFL have a history of scientific partnering, but it was not until 2011 that they established their innovation cell, Nestlé Institute of Health Sciences (NIHS), at EPFL to create a world-leading scientific facility for health science nutrition. The institute started with the vision of creating world-class excellence in biomedical research to better understand chronic human diseases and aging influenced by metabolism, genetics, and environment. Using teams composed of Nestlé researchers and researchers from EPFL, the institute performed fundamental research to combat the prevalence of complex chronic diseases that are increasing throughout the world. Nestlé has identified mega trends that if approached in an entrepreneurial way may become growth platforms for the company. For instance, according to the World Health Organization, up to 80% of cases of heart disease, strokes, and diabetes result from obesity. An estimated 1.46 billion adults worldwide are obese; 3 million die each year. Separately, Alzheimer's disease is a neurodegenerative disorder affecting 35 million people. Between 2000 and 2006, deaths related to Alzheimer's disease increased by approximately 46%. The rate of occurrence of Alzheimer's disease doubles every 5 years for those between 65 and 85 years of age. Nestlé's innovation cell addresses these "global diseases" by establishing itself at EPFL to conduct research on metabolic cognitive disorders and develop nutritional strategies to improve health. In addition, Nestlé works at EPFL in the development of innovative concepts and products at the interface between food and pharma that will benefit consumers' health worldwide.

The research collaboration with EPFL focuses on the areas of brain health, metabolic health, and gastrointestinal health. Chronic noncommunicable diseases such as obesity and diabetes are a result of a highly complex multifaceted relationship between genes, diet, and lifestyle that varies for each individual.

> Nestlé's objective to establish in EPFL is to reinforce our leading research capabilities such as integrated biology systems, next-generation sequencing, human genetics, metabonomics, and lipidomics. Targeted nutrition requires a strong underpinning of scientific proof as well as state-of-the-art diagnostics. We foresee that the collaboration between Nestlé and the EPFL will help reach our goal.
>
> —Head of the Nestlé Institute
> of Health Sciences

NIHS is fully integrated to Nestlé's Global Research & Development network. Hence, NIHS it is not consider a satellite organization but, rather, a main center for research with middle and upper management, which helps to overcome organizational barriers in order to transfer knowledge to the rest of the organization. Nestlé has a global network of R&D centers, and NIHS paves the way for a new approach to generate innovation within a university setting. The NIHS innovation cell anticipates further collaboration with other R&D corporate centers, such as their Product Technology Center in Konolfingen, Switzerland, focusing on infant formula, dairy products, and medical nutrition development, as well as their R&D center in Manesar, India, established in late 2011.

Nestlé's screening platforms integrating biochemistry with genomics-based systems add a pharmaceutical dimension to their innovation agenda and highlight the potential for this research to be translated into effective clinical commercial applications. This approach is expected to result in nutritional strategies to help slow and even prevent the onset of chronic diseases in ways that can be effective and more affordable than drugs. NIHS foresees that these disruptive innovation projects could be scaled by its business units.

> Nestlé has always led when developing new manufacturing processes, products, and packaging as well as new ways to connect with consumers. One of our biggest opportunities for growth is to focus research on science-based nutritional solutions to help prevent or manage chronic diseases. The work of the Nestlé Institute of Health Sciences is a new arm of our research capability. It will strengthen our position as the world's leading nutrition, health, and wellness company.
> —Chief Executive Officer, *Nestlé*

A common place for interactions between Nestlé's innovation cell and graduate students is the classrooms because some of the Nestlé employees are adjunct professors at EPFL. Regarding startups, Nestlé has had discussions with neighboring startups to develop techniques for genomics and protein analysis. In terms of culture, Nestlé believes that by being at the EPFL, the NIHS is being consistent with the global vision and corporate culture that embrace innovation.

Credit Suisse

In 2011, Credit Suisse IT group established an innovation cell at Innovation Square to develop and apply cutting-edge innovative technologies. A particular area of interest between Credit Suisse and EPFL is cloud computing architecture, in which data and programs run across a remote network of servers. This computing architecture is of particular interest to banks, which need to manage in an organized secure way large amounts of data. Several IT areas from Credit Suisse are present at the innovation cell, such as regulation reporting, document management, corporate web, workflow management, and infrastructure engineering. Each one of these IT areas has its own projects and works with the ecosystem around EPFL's School of Computer and Communication Sciences and Engineering. In some projects, they share development resources to make innovation development more efficient.

> The arrival of Credit Suisse at Innovation Square is a unique opportunity to bring together the worlds of banking and scientific research. The EPFL has developed numerous areas of expertise in the financial sector, not only in terms of IT infrastructure but also with its innovative financial engineering program and its training facilities for mathematicians.
> —Vice President for Innovation and Technology, *Transfer at the EPFL*

Credit Suisse's establishment at EPFL brought as an externality the influence of Credit Suisse's corporate culture. Due to its banking background, Credit Suisse has a risk-averse culture that, by being at EPFL, allowed Credit Suisse IT to adapt to a "Let's have a go" or "Ask for forgiveness, don't ask permission" approach. Knowing that not all initiatives will pay off, upper management has become more tolerant of exploratory projects that may lead to a dead end.

> Traditionally the banking industry tends to have a risk-averse culture and for the most part Credit Suisse has not been the exception. However, by being at EPFL we are exposed to a more entrepreneurial culture, and here we are more exploratory towards new projects knowing well in advance that not all initiatives will pay off.
> —IT Innovation Director, *Credit Suisse*

The Credit Suisse innovation cell at EPFL reconciled the corporate technology transfer resistance "pushbacks" by earning respect within the organization only after proving a few innovation success cases. This nascent track record established a reputation throughout the organization that generated more inquiries from the business units to initiate innovation IT projects. Even the New York office requested services to be developed by Credit Suisse IT at EPFL. Credit Suisse learned that in order to overcome the challenges related to transferring innovation to the rest of the organization, it is important to have a sponsor from a senior manager at Credit Suisse. The sponsor will not only be the internal costumer of the new innovation but also provide the budget to develop the idea.

Credit Suisse has been working on different research projects and initiatives with EPFL and is particularly interested in interacting more actively with master students to be able to attract promising new talent into the organization. Close to 50% of their employees at their innovation cell studied at EPFL. By being based on campus, the Credit Suisse innovation cell works on projects mainly with master students, allowing the innovation cell to screen out promising students who will fit with its culture.

> We find at EPFL an entrepreneurial academic environment. EPFL graduates represent a great recruitment pool for our growing team.
> —IT Innovation Manager, *Credit Suisse*

Credit Suisse is working with EPFL on different projects. These comprise data analysis and IT application laboratory data, including the application of compilation techniques and ideas from programming language theory for database query processing. Credit Suisse's IT innovation cell also works on generating massively parallel data management systems based on lightweight components. This project is referred to as Big Data, with the ultimate goal of handling and extracting information more efficiently. Another project of Credit Suisse and EPFL involves complexity theory and logic theoretical research on the foundations of query processing. This project specifically studies the complexity of database queries, which guides the development of efficient algorithms and helps to create an understanding of the fundamental limits to efficient query processing. Separately, the probabilistic database MayBMS project studies the management and processing of uncertain data and develops database management systems for such data. Related to data management on Web 2.0 is the Youtopia Project, in which EPFL designs declarative easy-to-use languages for specifying and solving coordination problems as they increasingly occur in social web applications. Finally, Credit Suisse also works with EPFL to make its servers more efficient in terms of power consumption for its data centers.

> Cloud computing is of particular interest to banks like Credit Suisse, which need large amounts of computing power. But this means that we have to rethink how to improve areas such as security and database organization, both aspects in which EPFL has strong competencies.
> —Dean of Computer Sciences at EPFL

More examples of collaborations between large companies and EPFL are listed in Figure 8.8. These include incubation programs, working with startups, internships for EPFL students, recruitment, joint research programs, and development of prototypes and business plans.

Company	Innovation Cell	Collaboration	Interest
Cisco Systems	• At EPFL Innovation Square since October 2010 • 35 Software developers and testers	• Cisco-EPFL strategic partnership including joint projects and research • Incubation programs and working with startups • Swiss federal project sponsorship (CTI)	• EPFL presentations and seminars from professors and Cisco • Internships at Cisco • Career days for new talent • Specific joint technologies and products: High-end router platforms, telepresence, IP cameras, digital technology, and desktop phones
Constellium	• Innovation Cells unit on campus since March 2008 • Cross-functional multidisciplinary team (commercial and technical) 7 employees	• Aerospace Global Marketing Team dedicated to marketing and innovation in aerospace industry	• Professor Endowment EPFL's Institute of Materials Science • Donation of 2.5 Mio CHF, co-funded with EPFL • Acceleration of ideas into prototypes and business plans
Nestlé Institute of Health Sciences	• Innovation Cell at EPFL since August 2011 with a mission to develop customized nutrition to prevent and cure diseases • Research based on genomics and metabolomics • Goal to recruit 200 people in a 10 year plan	• 500 million CHF investment in research • Joint research and PhD program collaborations • Collaboration with UNIL, CHUV and other Swiss academic partners	• Joint professors, researchers and scientists • Use of technical Platforms • Fields of research: Bioinformatics, Genomics, System Biology

Figure 8.8 Examples of collaborations at the Innovation Square Ecosystem.

DISCUSSION AND CONCLUSIONS

This chapter examined how innovative firms can get access and develop disruptive innovation by adapting their organizational structure through the setting up of an innovation cell physically separated from the rest of the organization to avoid the business unit's short-term vision. Moreover, innovation cells are to be located in a leading engineering university that strives for industrial liaison collaborations and technology transfer partnerships. By having an innovation cell at an engineering university campus, large firms are expected to secure novel technology and marketing insights as well as to be exposed to an exploratory culture likely influencing management to become more tolerant to risk and dead-end potential outcomes.

The host university needs to create a specialized innovation ecosystem in which large firms and research laboratories coexist. The proximity between firms and university fosters day-to-day interactions among different actors of the ecosystem, such as researchers, students, startups, venture capitalists, government entrepreneurial agencies, and other large companies. The new ideas that result from these interactions can be transferred to the rest of the firm to be developed into new products and services. However, physical proximity by itself does not guarantee innovative breakthroughs. The host university has to offer a dedicated location as well as a team that provides services to its tenants, facilitating and managing formal and informal interactions. In turn, companies need to establish internal efficient organizational channels that can absorb the outcome of their collaboration with external partners (Cohen, Nelson, & Walsh, 2002; Teece, Pisano, & Shuen, 1997).

We observed that in order to generate disruptive innovation, the role of companies' innovation cell is different from that of their fundamental R&D. The participation and involvement as the leading innovation unit for a particular project depend on the degree of disruptiveness and the technology development stage (Keil, 2002). Firms lean on R&D for exploratory early stage initiatives, whereas they rely on their business units' development capabilities when technology has been proven and prototypes exist. It is only when a project is considered to have a high level of disruptiveness in which R&D and business units are not capable of developing the product themselves that the innovation cell gets involved and leads the initiative. This organizational framework suggests that centralized organizational structures are suitable for disruptive innovation, whereas decentralized organizations at the business unit level are suitable when developing incremental innovation.

In summary, innovation cells leverage existing R&D capabilities with new ideas and technologies from the university ecosystem. Innovation cells also network through specific technological events and scientific conferences, participate at intercompany social events, and set up collaborative projects with different players of the ecosystem.

The innovation cell needs to have a multidisciplinary team in which commercial–business development and technical–scientific experts work together to deliver market-ready prototypes and innovative business plans in short time frames. The innovation cell should have the right people with the right mindset and organizational clout. It should include young talented engineers and senior managers with a successful track record of launching new products capable of influencing business units and headquarters and securing funding to undertake innovation initiatives.

Innovation cells have an adaptable organizational structure and culture conducive to the establishment and maintenance of cooperative partnerships with the different actors of the innovation ecosystem (Ahuja & Lampert, 2001; (Dushnitsky & Lenox, 2006; Guth & Ginsberg, 1990). For companies operating in such environments, this can represent a source of competitive advantage to generate disruptive innovation. The observed formal and informal interactions and evidence of disruptive innovation developments among the large companies and other actors of the EPFL Innovation Square ecosystem reinforce the notion that partnership collaboration found in an open innovation ecosystem can lead to the development of disruptive innovation that can turn into profitable new product growth platforms (Avnimelech & Teubal, 2006; Leamon & Lerner, 2012).

ACKNOWLEDGMENTS

This chapter was developed as part of the doctoral thesis of Felix Cardenas under the supervision of Professors Tony Davila and Daniel Oyon. The chapter was made possible through the support of the Swiss National Science Foundation (SNSF), National Council of Science and Technology (CONACYT), the University of Texas at Austin, and the Instituto Tecnologico y de Estudios Superiores de Monterrey.

Notes

1. Chesbrough's (2002b) definition of open innovation is "a paradigm that assumes that firms can and should use external ideas as well as internal ideas, and internal and external paths to market, as they look to advance their technology. Open innovation processes combine internal and external ideas into architectures and systems."
2. According to the World Economic Forum's *Global Competitiveness Report 2012–2013*, Switzerland is ranked the top country in terms of competitiveness due mainly to its strong performance related to innovation. Switzerland's scientific research institutions are among the world's best, and they have a strong collaboration between academia and business sectors. In addition, there is high spending on R&D by companies, which creates an environment prone to generate marketable products with strong intellectual property protection. Out of 144 countries, Switzerland ranks number one for company spending on R&D and university–industry collaboration in R&D. It ranks number two in capacity for innovation; quality of scientific research institutions; and patents, applications per million inhabitants.
3. The 2014–2015 *Times Higher Education World University Rankings*' Engineering and Technology table positions both ETH and EPFL among the top 12 engineering and technology universities in the world, along with US and UK universities such as MIT, Caltech, Stanford, Oxford, and Cambridge. The ranking of the world's top 100 universities for engineering and technology employs 13 carefully calibrated performance indicators to provide the most comprehensive and balanced comparisons available, which are trusted by students, academics, university leaders, industry, and governments.
4. Similar to that found at Building 20 at the MIT. Building 20 was a temporary wooden building constructed during World War II at MIT housing the radiation laboratory, which conducted research in electronics, electromagnetism, microwave physics, and communications. It is believed that the high innovation output of Building 20 was due to the day-to-day interaction with diverse multidisciplinary scientists working in proximity to each other. At one time, more than 20% of the physicists in the United States, including nine Nobel Prize winners, worked in that building. After the radiation laboratory ceased operations,

Building 20 continued serving as an incubator for small MIT programs, research, and student entrepreneurship activities before it was demolished in 1998.

References

Afuah, A. (2003). *Innovation management: Strategies, implementation and profits.* New York: Oxford University Press.

Ahuja, G., & Lampert, C. M. (2001). Entrepreneurship in the large corporation: A longitudinal study of how established firms create breakthrough inventions. *Strategic Management Journal, 22*(6-7), 521-543.

Avnimelech, G., & Teubal, M. (2006). *Evolutionary innovation and technology policy: A four-phase high-tech policy model.* Paper presented at the DRUID Summer Conference.

Chesbrough, H. (2002a). Graceful exits and missed opportunities: Xerox's management of its technology spin-off organizations. *Business History Review, 76*(4), 803.

Chesbrough, H. (2002b). Markets for technology: The economics of innovation and corporate strategy [Book Review]. *Journal of Economic Literature, 40*(4), 1275-1276.

Chesbrough, H., & Rosenbloom, R. S. (2002). The role of the business model in capturing value from innovation: Evidence from Xerox Corporation's technology spin-off companies. *Industrial & Corporate Change, 11*(3), 529-555.

Chesbrough, H. W., & Tachau, J. (2002). *Innovating an outsourced R&D process for Matsushita Electric (MEI): Launching the Panasonic Digital Concepts Center.* Harvard Business School Case 9-602.

Cohen, W. M., Goto, A., Nagata, A., Nelson, R. R., & Walsh, J. P. (2002). R&D spillovers, patents and the incentives to innovate in Japan and the United States. *Research Policy, 31*(8-9), 1349.

Cohen, W. M., & Levinthal, D. A. (1990). Absorptive capacity: A new perspective on learning and innovation. *Administrative Science Quarterly, 35*(1), 128-152.

Cohen, W. M., Nelson, R. R., & Walsh, J. P. (2002). Links and impacts: The influence of public research on industrial R&D. *Management Science, 48*(1), 1-23.

Cooper, R. G. (2003). Your NPD portfolio may be harmful to your business's health. *Research Technology Management, 47*(1), 31-43.

Davila, T., Epstein, M., & Shelton, R. (2005). Making innovation work: How to manage it, measure it, and profit from it [Book Review]. *CIO Insight 57*

Davila, T., Epstein, M., & Shelton, R. (2006). *Making innovation work: How to manage it, measure it, and profit from it.* Upper Saddle River, NJ: Wharton Publishing.

Day, G. S. (2007, December). Is it real? Can we win? Is it worth doing? *Harvard Business Review.*

Duchesneau, T. D., Cohn, S. F., & Dutton, J. E. (1979). *A study of innovation in manufacturing: Determinants, processes, and methodological issues*: Orono, ME: Social Science Research Institute, University of Maine.

Dushnitsky, G., & Lenox, M. J. (2005). When do incumbents learn from entrepreneurial ventures? Corporate venture capital and investing firm innovation rates. *Research Policy, 34*(5), 615-639.

Dushnitsky, G., & Lenox, M. J. (2006). When does corporate venture capital investment create firm value? *Journal of Business Venturing, 21*(6), 753–772.

Ettlie, J. E. (1983). Organizational policy and innovation among suppliers to the food processing sector. *Academy of Management Journal, 26*(1), 27–44.

Griliches, Z. (1979). Issues in assessing the contribution of research and development to productivity growth. *Bell Journal of Economics, 10*(1), 92–116.

Guth, W. D., & Ginsberg, A. (1990). Guest editors' introduction: Corporate entrepreneurship. *Strategic Management Journal, 11*(4), 5–15.

Hall, B. H. (1992). *Investment and research and development at the firm level: Does the source of financing matter?* Cambridge, MA: National Bureau of Economic Research.

Himmelberg, C. P., & Petersen, B. C. (1994). R&D and internal finance: A panel study of small firms in high-tech industries. *Review of Economics & Statistics, 76*(1), 38.

Keil, T. (2002). *External corporate venturing: Strategic renewal in rapidly changing industries*. Westport, CT: Quorum.

Kim, W. C., & Mauborgne, R. (1999). Strategy, value innovation, and the knowledge economy. *Sloan Management Review, 40*(3), 41–54.

Leamon, A., & Lerner, J. (2012). *Creating a venture ecosystem in Brazil: FINEP's INOVAR Project*: Cambridge, MA: Harvard Business School.

Lerner, J., Schoar, A., & Wongsunwai, W. (2007). Smart institutions, foolish choices: The limited partner performance puzzle. *Journal of Finance, 62*(2), 731–764.

Markides, C., & Geroski, P. (2005). *Fast second: How smart companies bypass radical innovation to enter and dominate new markets* (Vol. 178). San Francisco: Jossey-Bass.

Markides, C., & Oyon, D. (2000). Changing the strategy at Nespresso: An interview with former CEO Jean-Paul Gaillard. *European Management Journal, 18*, 296–301.

Markman, A. (2012, December 4). How to create an innovation ecosystem. *Harvard Business Review*.

Munson, F. C., & Pelz, D. C. (1979). *The innovating process: A conceptual framework* [Working Paper]. Ann Arbor, MI: University of Michigan.

O'Connor, G. C., & DeMartino, R. (2006). Organizing for radical innovation: An exploratory study of the structural aspects of RI management systems in large established firms. *Journal of Product Innovation Management, 23*(6), 475–497.

Rozanov, A. (2005). Who holds the wealth of nations. *Central Banking Journal, 15*(4), 52–57.

Sala-i-Martín, X., Bilbao-Osorio, B., Blanke, J., Crotti, R., Hanouz, M. D., Geiger, T., & Ko, C. (2012). The Global Competitiveness Index 2012–2013: Strengthening recovery by raising productivity. In *The Global Competitiveness Report 2012–2013* (pp. 49–68). Geneva, Switzerland: World Economic Forum.

Schumpeter, J. A. (1934). *The theory of economic development: An inquiry into profits, capital, credit, interest and the business cycle* (R. Opie, Trans.). Cambridge, MA: Harvard University Press.

Teece, D. J., Pisano, G., & Shuen, A. (1997). Dynamic capabilities and strategic management. *Strategic Management Journal, 18*(7), 509–533.

9

Building an Innovation Coral Reef

The Austin Technology Incubator Case Study

**GREGORY P. POGUE, KEELA THOMSON,
ROSEMARY FRENCH, FRANCESCA LORENZINI,
AND ARTHUR B. MARKMAN** ∎

INNOVATION IN ORGANIZATIONAL SETTINGS

This book is focused on open innovation. For innovation to be open, many different people have to get involved, particularly those who might not talk with each other normally. As a result, it is crucial to design social structures that promote interactions to support innovation.

What should those social structures look like? In order to address this question, we explore the social structures that are effective for supporting new entrepreneurial ventures. Successful technology incubators, accelerators, and communities do a good job of bringing together people who can allow new businesses to thrive. We can learn much about successful innovation by first understanding the communities in which new businesses grow.

In this chapter, we discuss the concept of innovation and some of the barriers to innovating successfully. Then, we discuss the social structures that support new businesses and introduce the concept of the coral reef, which is a community structure that we have found to be successful. Finally, we discuss ways to implement a coral reef within an organization to support innovation.

BARRIERS TO INNOVATION

There are many definitions of innovation, and so it is important to be clear about what we mean by this concept at the outset. We assume that innovation, the process of turning novel ideas into tangible goods and services, is both a risky and a powerfully disruptive process and thus quite fragile in nature. Although innovation involves high risk in the early stages of product development, it also has the capacity to improve quality of life and positively impact the economic situations of individuals, institutions, and regions.

Idea-to-product development occurs in existing firms, both large and small, as well as within newly established startup companies. Each of these two settings has its own host of unique challenges. Startups often form around an innovative idea, and so the process of generating ideas has already been dealt with. However, startups have to create an infrastructure (including an appropriate leadership and operational team), find outside mentorship, deploy a business model, and develop a strategy for entering one or more markets.

Large firms often have an infrastructure in place for marketing and distributing products, but they often do not have mechanisms in place to support the generation of new ideas. In addition, managers are often unwilling to take on new and highly innovative projects because these types of projects have a high risk of failure. Failure can have a significant negative impact on both their compensation and their reputation within the organization. Although innovation has a high probability of failure in both venues, startups are often organized with the knowledge that there is some chance they will fail. In contrast, large organizations are often less tolerant of failure.

Overcoming these barriers requires both strategies for effective innovation and also appropriate personnel and mechanisms for supporting the kinds of collaboration that promote the components of innovation to realize products and benefits to organizations and society.

In this chapter, we are particularly interested in the social component of innovation. We want to understand the strategies that allow firms to bring people into contact to help them develop new ideas successfully and to generate plans that allow resources to be put behind these ideas in order to drive them forward toward the market.

It is crucial to overcome these barriers because successful innovation leads to improved medical procedures, alternative energy sources, and better ways to manage data, and it drives the growth of companies and economies alike. Innovation has the potential to improve quality of life, create new job opportunities, and make a positive impact on regional economies.

In order to develop lessons for the way large organizations can structure their social environment to support innovation, we start with an extended exploration of the innovation environment, or ecosystem, for small firms or ideas within larger organizations. In particular, we have explored the ecosystems that surround entrepreneurial communities. In order to place this work into perspective, we first examine some of the core challenges surrounding innovation and then discuss previous theoretical approaches to understanding the ecosystem surrounding small firms. We are not the first to explore the structure, components, and actions of innovation ecosystems, so we seek to set the context for open innovation's relations to ecosystem thinking.

THE TRANSITION FROM CLOSED TO OPEN INNOVATION

Through the first half of the 20th century, innovation happened largely within the context of single firms. It was common for large firms to orchestrate innovation projects, for which they would acquire all the financial, technological, and human resources necessary to develop a new technology and commercialize it through a product vehicle. The large industry leaders with access to excellent resources, such as Bell Labs, GE, DuPont, and Merck, performed the most internal research and development (R&D) and subsequently reaped the benefits (Chesbrough, 2003). Relatively recently, however, large firms have lost their lead in innovation.

Today, there are better, more economical ways of reaching the market without investing in costly in-house research facilities. The innovation process has been democratized in the past 40 years—with the rise of the applied conceptual expertise within the research university, the availability of risk financing, and the advent of the Internet era. The passage of the Bayh–Dole Act in 1980 paved the way for private and public research institutions to leverage public research funds to make fundamental scientific discoveries and to transfer technology to the private sector, in which technologies are transformed into useful and valuable innovative products and services. Growth in capital to finance early stage ventures from business angels, venture capital organizations, and corporate-sponsored research allows early stage entrepreneurs to take creative ideas (which are the raw material of innovation) and turn them into products, allowing for financial reward to be shared by inventor and investor alike.

In response to the growing dominance of small firms as engines of innovation, large companies have tried to become more agile by looking outside their four walls to other entities and using a more "open" strategy for innovation—integrating parts from disparate partners into a single marketable product. Thus, in the 21st century, innovation has evolved from a focus on large, tightly controlled projects toward a more dynamic process driven increasingly by smaller and newer companies and individual research laboratories (Chesbrough, 2003, 2006).

In this new open paradigm, innovation projects are not controlled by a single company or group but are cultivated within an ecosystem of "role" players: The team that originates an innovative idea often is not the same as the one that funds its development, finalizes development, or commercializes it. In addition, as companies grow and evolve, they may find that improvements to existing products as well as new ideas originate from external groups. For example, although Apple clearly developed the iPhone and iPad platforms, the success of these platforms rests on an open innovation model that allows external developers to create compatible applications.

However, as technology and the ways that it is used evolve, innovation is becoming more difficult in some ways. As technology progresses, the resources and skills needed to innovate or transform basic scientific discovery into a useful and profit-bearing product are becoming ever more advanced in many fields. People require more training to understand the current technologies, the ruling regulatory policies, current market needs, and strategy for new product introduction. Clearly, innovation projects do not cleanly fit into one discipline (Chesbrough, 2003, 2006) and thus may require a cross-disciplinary team with many disparate skills (Bunderson, 2003; Stoker, Looise, Fisscher, & DeJong, 2001). The growth in team complexity creates a more dynamic innovation process in which the required human resources might change dramatically within a company over the life span of developing a single technology.

How are these teams of multidisciplinary experts identified, incentivized, and brought into a team to develop an idea into a product using open strategies? The answer largely lies in the activities occurring in the innovation ecosystems surrounding the innovators and associated "role" players.

THE HISTORY OF INNOVATION ECOSYSTEM MODELS

The open innovation concept discussed previously is only one of the many models proposed to conceptualize and categorize the ways in which new technology development is successfully performed. Here, we discuss the history of innovation theory and the ways in which these concepts have evolved in response to market realities.

In the early 1800s, as the Industrial Revolution began to spread to other countries outside of Great Britain and manufacturing became increasingly central to many economies, various economists cited a need for mechanisms to foster early stage business development. One of the earliest conceptual models of innovation was proposed by the German economist George Friedrich List in 1850, in which he argued that Germany needed a national system in order to catch up with Great Britain (Andersen, 2011). List successfully developed a national manufacturing base to protect nascent industries from foreign competition until they attained the ability to compete on the international level. Other developing nationals at the time were inspired to follow List's model, such as Japan (Andersen, 2011). This concept is analogous to the business incubation concept, in which early stage companies are nurtured in a close-knit scenario in order to improve their potential for success.

Proximity of like-minded businesses has been proposed in a number of models as a key to success. In 1899, the English economist Alfred Marshall coined the term "agglomeration" and advocated the advantages for companies to be located in proximity to other companies in the same industry. This concept was expanded to include the role of entrepreneurs and capital in 1950 when the Swedish economist Eric Dahmén proposed the need for "development blocks" as a mechanism for Sweden's economic development and industrial makeover (Andersen, 2011).

Furthermore, the importance of the advantages inherent in specific geographic regions has also been highlighted in a number of innovation models. In the late 1980s, Christopher Freeman and Bengt Ake Lundvall coined the concept of a "national system of innovation" (Freeman & Lundvall, 1988; see also Freeman, 1995). This model exposed not only the need for the key players of entrepreneurs, companies, and capital but also the effect of governmental policy on the labor market, education, and other economic development policies.

Among others, Smilor, Gibson, and Kozmetsky (1988) analyzed the various players in the evolving innovation ecosystem within Austin, Texas, and coined the term "technopolis." In their article, the authors present seven segments within the technopolis: the university, large technology companies, small technology companies, the federal government, state government, local government, and support groups. Key findings in the study include the pivotal role of the university, the need for continuity in government policies, the catalytic role of large corporations, the importance of new company creation, and the need for consensus for the sustained development of the technopolis. In summary, a coordinated approach is needed for emerging high-technology company development.

In 1990, Michael Porter published *The Competitive Advantage of Nations* (Porter, 1990a; 1990b) that introduces the concept of "clusters" as a vehicle for economic development of industries, regions, and nations. Although this cluster analogy contained many of the same elements of innovation models mentioned previously, Porter's model achieved a higher level of recognition than many of the other previous models. Today, many regions and/or nations have deployed aspects of Porter's clustering concept into their policy (Muro & Katz, 2010).

More recent models have emphasized the interplay between various groups. In the 1990s, Loet Leydesdorff from the University of Amsterdam proposed a "triple-helix" model of innovation (Etzkowitz & Leydesdorff, 1995), in which regional universities, governments, and firms all play a role in the promotion of sustainable technology development. Building on this concept, the quadruple helix model proposes four interrelated sectors: public, private, civil, and academic. Both of these models emphasize the role that governments and universities play and their effects on regional innovation, and they reflect derivations of the technopolis model (Smilor et al., 1988).

Studies suggest that localized hubs of institutions, businesses, and personnel promote successful innovation (Muro & Katz, 2010). These clusters are often specific in terms of geography as well as market sector. Silicon Valley is the most famous example of such clusters. In such clusters, there are frequent interactions between technology experts, entrepreneurs, funding partners, and other associated members of the tech community, on both a formal and an informal basis. A variety of models have attempted to re-create Silicon Valley in the form of region-specific and even shared working space-specific hubs such as technology incubators and science parks. Many of these clusters are built around a university site that provides access to novel research and talent. Unfortunately, billions of dollars have been spent to reproduce Silicon Valley in distant locations with little return (Wadhwa, 2013). The unique milieu of university, industry, capital, and the quality of people involved has proven difficult to replicate.

One reason why regions such as Silicon Valley may be difficult to re-create is that there are interesting sociological factors that are difficult to legislate. For example, Saxenian (1996) compares the development of the technology communities in the Boston area and in Silicon Valley in the 1970s and 1980s. She argues that these regions had similar government and university support for innovation, but they differed in the size of the firms in the regions. Boston had large technology firms, whereas Silicon Valley was dominated by startups. The large firms in Boston tended to punish executives who led failed projects. In contrast, executives who led failed startups in Silicon Valley were quickly pulled back into new projects. Because the business talent in Silicon Valley was able to learn from its mistakes and to take risks without fearing permanent career damage from failure, West Coast firms ultimately tried out new technologies at a higher rate than East Coast firms. Ultimately, that led to the dominance of Silicon Valley over Boston. It is difficult to re-create this kind of environment, however, because much depends on the people making up the community and not just the legal and economic climate of the region.

Likewise, studies show mixed results in terms of whether business incubation promotes successful innovation throughout the world. Clustering and incubation appear to work better in certain geographic regions than in others. Although clustering is strongly indicated as a means to promote regional economic growth, more research needs to be performed with regard to the reasons why some geographies have higher innovation success rates than others. Such factors as cultural barriers, governmental policy, and transportation infrastructure have been proposed as possible factors involved in the variation in success rates.

In addition to the conceptual models behind novel product development, a number of theories have been proposed regarding the ways in which business incubators can nurture and improve the outcomes for those ideas. Hansen, Chesbrough, Nohria, and Sull (2000) propose the development of the networked incubator, in which the value proposition is not found in space, coaching, services, or traditional activities but, rather, by the purposeful and organized activities of networking startups to funding and partners to accelerate their business. The transition from financial capital emphasis to intellectual and social capital is key to providing more economic opportunities for the incubated companies.

OPEN INNOVATION AS AN ECOSYSTEM

As attractive as networked incubators appear, businesses need more than an incubator and nascent relationships. Whereas Silicon Valley offers a ready-made ground for networking and the realization of potential of intellectual and social capital, most regions are not well prepared for this model. Indeed, most regions lack mechanisms to attract and organize the necessary talent to provide the scale and scope to reduce operational costs through shared resources; the stimulation to encourage individuals to pursue high-value and risky business options; and mechanisms to exploit proximity to develop partnerships for advice, strategy, team building, and commercial acceleration. Therefore, a broader, more resource-agnostic model is required to build a translatable and operational model for new venture acceleration to developing or disadvantaged regions.

Our research has explored technology incubators in order to help us understand how ecosystems can function. As discussed in the next section, technology incubators were initially designed to provide nascent startups with office space. Throughout the years, they have developed to support a broad set of interactions among startups, funders, business talent, technology experts, and members of the government and university communities.

In the next section, we describe our work with the Austin Technology Incubator (ATI) at The University of Texas at Austin, which is one of the premier business incubators in the United States (Wiggins & Gibson, 2003). Our observations have led us to characterize the ecosystem surrounding ATI as similar to that of a coral reef. In the rest of this chapter, we explore how this "coral reef" model overcomes many of the cited barriers to incubation. The coral reef model posits that an innovation ecosystem, like a coral reef, should provide a dense centralized hub of resources and interaction.

ATI MODEL

Business incubators have been supporting the growth of innovation in new and small companies for decades. Since the first incubator was established in Batavia, New York, in 1959 (Smilor & Gill, 1986; Wiggins & Gibson, 2003), business incubation models have been evolving to meet the needs of the startups they serve. Early business incubators tended to provide cheap, shared office space to help lower startups' costs. Over time, many have shifted their focus toward providing value-adding services such as coaching and advice, including "networked incubators" that provide "value-adding" services as other incubators but also supply strategic access to talent and capital (Aldrich & Zimmer, 1986; Bollingtoft & Ulhoi, 2003; Grimaldi & Grandi, 2005; Hansen et al., 2000; Ratinho, 2011). Business incubator function and practice have evolved from more tangible offerings to more intangible contributions that reduce business risk and accelerate commercialization.

ATI is an example of a networked incubator that has created its own innovation ecosystem. Founded in 1989 as a part of The University of Texas at Austin (UT), ATI was born from the first organized model for regional innovation ecosystem building—the technopolis—and benefits from the vision of providing entrepreneurs with close proximity to wisdom, services, and capital providers (Smilor & Gill, 1986; Wiggins & Gibson, 2003). ATI assists early stage, technology-based firms operating in one of four industry verticals—health care, clean energy, information technology, and wireless—as well as a portfolio of student companies from UT. Reputation, recommendation, and ATI events are the most frequent manner that companies become aware of ATI's services and potential value. Companies apply through interactions with ATI directors of the appropriate vertical and are evaluated for "fit" into ATI. The criteria for company acceptance is not due solely to the potential market value of the product or service but, rather, a measurement of the "fit" of the company's leadership team into the local ecosystem. This ecosystem provides legal and accounting services at below-market prices or in exchange for equity, market channels, technology expertise, developmental experts, talent for corporate activities, and funding sources. In order to be accepted, firms must go through a formal competitive application process that screens for the viability of the technology, the market opportunity, the credibility and "coachability" of the company's team, the fit with one of the ATI verticals, and the ability of ATI to provide value to the company in terms of commercial acceleration or capital access.

One of the most important contributions that ATI makes to companies is that it surrounds them with talent that they cannot yet afford. The directors at ATI are industry experts who provide extensive coaching services and also spend much of their time maintaining an informal external network of knowledgeable community members such as funding providers, industry experts, large company executives, technological experts, lawyers, and other people who can be helpful to the member companies. The final stage in the application process for a company is giving a pitch in front of a "success committee" composed of ATI staff and a select group of relevant "externals" selected specifically for their ability to assist the company to grow and provide relevant feedback to ATI. The recommendations of the success committee are in essence a two-way commitment: the company commits to the incubator, and the incubator and externals commit to helping the company find the capital and market access it needs for success. Once admitted to ATI, companies are often referred to externals both to help them obtain deals with funding providers or customers and to receive advice.

Member companies at ATI have a clear goal or return on investment (time, effort, and financial) model: to obtain seed round or Series A funding. Much of the assistance that ATI provides is centered around developing companies to the point at which they are ready to compete for funding and helping companies find and pursue opportunities to obtain funding from grants and equity investments by angel networks and venture capitalist firms. Generally, companies aim to obtain between $500,000 and

$5 million in funding. Once companies have obtained Series A funding, they typically graduate from the ATI program.

OBSERVATIONS AT ATI: ATI IS LIKE A CORAL REEF

Many of the practices at ATI provide a model of innovation ecosystems that is similar to a coral reef in several ways (Table 9.1). Coral reefs represent unique ecosystems within the broader ocean environment. Although marine life is present between the beach or coast and the reef, it is dilute and not organized into a relational environment. Life beyond the reef dwells in a deeper water context that is rich with life, including larger, more predatory organisms live, thus requiring "schooling" for community, not individual, protection and the primary relationship transactions are in the currency of food and nourishment. The reef stands in stark relief of these adjacent environments. It is a relationship-rich, localized

Table 9.1 ATI AND CORAL REEF STRUCTURE AND CONTENT COMPARISON

Attribute	Coral Reef	Austin Technology Incubator
Size	Occupies 0.01% of ocean floor surface.	Occupies 0.1% of the square footage at UT Austin in part of one floor of one building.
Diversity	32 of the 34 animal phyla are found in coral reefs, compared with only 9 in rain forests; virtually all phyla of plants are found in reefs.	ATI unites university researchers, students, new entrepreneurs, experienced entrepreneurs, expert generalists, service providers, capital providers, etc.
Specialized conditions for growth	Found in ocean depths of less than 150 feet and temperature range of 23–29°C and salinity range of 32–40%.	Collaborative conditions found in Austin, not larger cities or neighboring states. Established reputation for fair trading of knowledge, talent, and resources and proven track record of delivering value.

SOURCE: DATA for coral reef from Henkel (2010) and Swart (2013).

environment with organisms aggregate dilute chemical nutrients to provide the basis for extraordinary life support. Reefs contain many organisms that trade in mutualistic biological currencies, including shelter, protection, nutrient processing, and cleaning. Furthermore, the reef is a fragile environment, which can be easily disrupted by changes in the environment, nutrient flow, and population dynamics of its inhabitants.

ATI is likewise a very small part of UT Austin, occupying part of one floor of one building among the ~230 buildings on the campus. Participants in entrepreneurship tend to be scattered throughout a city based on various factors. However, within ATI, a high concentration of stakeholders critical to startup survival frequent and make particular investments in ATI-incubated companies to increase their chances of survival and funding. ATI advice supports companies as they seek to develop technology into products, develop the team that can both lead and function as a unit, identify and penetrate initial market opportunities, and develop a scalable organization strategy for growth. ATI further protects its companies from the predatory activity of some organization through advice and wisdom networks. This environment is fragile like a reef, and ATI must maintain the quality of relationships, participants, and value transactions to keep all participants fair and incentivized. Next, we explore this analogy between ATI and a coral reef in more detail.

Although coral reefs occupy only a very small surface area in comparison to the ocean as a whole, they are home to the densest population of life on earth. Similarly, ATI creates a dense community with a critical mass of concentrated talent. The directors at ATI network to assemble the contributors necessary to nourish startups in one place, such as advice, knowledge, investors, and recommendations to service providers such as lawyers and consultants. Furthermore, ATI attracts expert generalists who have broad and significant knowledge about commercialization and company operations. This usually comes through serial entrepreneurial activities and/or deep experience with capitalization strategies and company operations. They provide cross-cutting expertise that links sector-specific advice into a strategic whole. This amalgamation of expertise that a startup company needs is accomplished through the dedicated networking efforts of the ATI staff over an extended period of time and also through the favorable conditions of the city of Austin. Austin is a hub for technology startups and contains many of the right resources. It is home to several venture firms, as well as the Central Texas Angel Network, which is one of the most active angel networks in the country. There are also many technical experts due to the proximity of UT Austin and the technology hub in Austin as well as a healthy population of serial entrepreneurs. This concentration of social capital lowers the transactions costs to find the right talent.

As in a coral reef, ATI is able to leverage this talent by creating an environment of symbiotic or mutually beneficial relationships (Table 9.2). The company members benefit greatly from advice and funding received, and the externals benefit as well. ATI's reputation as an effective incubator helps it attract the most promising startups in the region. Many externals volunteer time at ATI in order to gain the opportunity to search for their next investment or employment opportunity. Others are entrepreneurs who have already achieved significant exits and want opportunities to remain connected to their industry and give back to the community. This emphasis on relationship building not only promotes startup success but also encourages serial entrepreneurship. The development of these high-quality business relationships creates an atmosphere in which "externals" want to be involved with multiple ATI companies. In turn, as member company entrepreneurs see the value of these relationships, many founders return to ATI with new business ideas.

The specific selection criteria and defined member company objectives ensure that there is a tightly integrated community organized around a central goal. Coral reefs shelter and nourish specific types of organisms from the more harsh conditions outside this protected environment. Similarly, in its admissions process, ATI seeks out companies that will thrive in the environment it provides and align with its return on investment model. Furthermore, companies are expected to provide valuable additions to the environment. Not all ocean life finds a home in a reef. Those participating must function in concert with the established biological niches and mutually beneficial relationship necessitated by close

Table 9.2 COMPARISON OF CORAL REEF AND ATI FUNCTION

Functionality	Coral Reef	Austin Technology Incubator
Houses diverse and robust networks	Densest population of plants and animals on earth.	Robust, diverse regional networks around specific market sectors (wisdom, capital, entrepreneurs, technologies, expert generalists).
Encourages mutually beneficial relationships	Corals, coralline algae, waves, fish, sea urchins, and sponges all contribute to and strengthen the reef structure and ecosystem.	All participants agree on and support return on investment model—funding for rapid growth companies.
Trades in various forms of "currency"	Food, protection, mutualistic relationships, cleaning, housing, nutrient filtration, and micronutrient metabolism.	Employment opportunities, "giving back" to community, investment and service opportunities, being "where the action is."
Efficiently recycles resources through open systems	Phytoplankton, seaweed, and algae filter nutrients to corals and are eaten by fish and crustaceans, which pass nutrients through the food web.	Entrepreneurs, advisers, and graduating students "graduate" from ATI and then return to reinvest in companies and mission.
Colonizes new space based on critical mass in a local network	Horizontal and vertical growth in warm, shallow, clear, and agitated waters with the correct nutrient recycling balance.	Serial introduction of technology "verticals" based on fit and value to Austin community: wireless, information technology, clean energy, health care, and student-based entrepreneurship.

proximity. Similarly, ATI is home to companies that mesh well with the expertise and value that it can provide, and it specifically screens out companies that do not fit into one of its verticals or that are not actively seeking early stage funding. Those companies or entrepreneurs who seek to act as "lone rangers" or are unwilling to take advice or direction from the more experienced leadership to whom ATI exposes them need to find incubation from other sources. For example, in the success committee process at ATI, the community is asked to determine whether a prospective member company can be assisted by ATI and is a good fit in the Austin ecosystem. If a company's leader would add value to the larger entrepreneurial community or has already proven to be a valuable member of the network, the company might be given a boost in the admissions process.

The ATI network is open and informal, which allows for a healthy flow of people and ideas into and out of the network. As in a coral reef, there is no clear boundary line between ATI and the rest of the community. People from many organizations are free to mingle, and often talent is recycled through ATI in many different roles. Indeed, the rapidly evolving and innovative climate of ATI draws in expertise from throughout the region and nation using ATI as an organizing center for communication and community. For example, the founding team of a company that has already graduated from ATI might be called upon to give advice to a new team working in the same industry.

Finally, the nonprofit nature and university affiliation of ATI allow for an extraordinarily diverse network. The reef works to grow and build its environment, but not at the expense of its constituents. Mutual benefit of the individual players and the reef as a whole is necessary. ATI does not stand to significantly gain directly from the participation of network members, and it is able to stand as a neutral party and provide a "safe zone" for both companies and advisors. This position increases the credibility of ATI's recommendations to important stakeholders and capital providers. These trust networks are jealously maintained for clarity, quality, and fluidity of conversation.

The outcomes from these structural and relational activities at the ATI reef are impressive and have catalyzed a remarkable change in the Austin community and the central Texas region. Since its inception in 1989, ATI has helped companies raise more than $1 billion in early stage capital and produce more than $1.5 billion in revenue, launch six initial public offerings on the NASDAQ exchange, create more than 10,000 local jobs, and facilitate dozens of acquisitions generating immense wealth in Austin. Since the "Great Recession" of 2008, ATI's reef strategy has been most keenly observed where it has done the following:

- Worked/trained 70 companies (admitted ~1/15 applicants)
- Found funding for 85% of resident companies
- Facilitated the raising of more than $500 million in investor capital
- Saw greater than $500 million in local company exits
- Created more than 400 jobs/year, direct and indirect
- Returned $67 for every $1 invested in ATI by the city of Austin
- Produced greater than $880 million in local economic impact in a 10-year period (Jarrett & Field, 2014)

These outcomes speak to the effectiveness of using the reef model to coalesce a critical mass of resources to drive value for both resident companies and the region under very challenging economic conditions in which other organizations failed to produce such a return on investment.

THE REEF MODEL SOLVES INNOVATION ECOSYSTEM BARRIERS

The elements of the ATI reef model combine to overcome barriers to the commercialization of innovations by mitigating some of the risk inherent in early stage technologies and lowering the costs of obtaining the necessary resources. ATI provides a concentration of social capital targeted toward specific industries and at the appropriate stage of innovation, providing insight into needs and potential value of each member of the network. The directors provide an informal matchmaking service to appropriate high-value individuals that lowers the transactions costs to find the right talent. The community is naturally organized around early stage innovation in specific industry sectors, which allows it to be more focused and self-organizing for full value provision. Members of this community mutually benefit from interactions which creates an incentive to participate in start up operations, link with funders or provide mentoring. The informal nature of the network makes it flexible and dynamic, allowing for a healthy flow of people and ideas. Finally, ATI is viewed as a "safe zone" with no profit motive.

The startup model itself reduces barriers to innovation faced by large tech-based corporations. A startup is free from many hurdles to developing an innovative product found in existing corporations, including decision-making and budgetary bureaucracy, measures and management of risk, and demands for quarterly revenue reporting (Chesbrough, 2003). A lean startup can de-risk novel product concepts. This in turn makes the innovation more attractive to corporate parties that may potentially license the innovation or acquire the startup.

RECOMMENDATIONS FOR CREATING CORAL REEFS

These observations at ATI can be adapted to creating innovation ecosystems more generally. Regions, universities, and companies should seek the necessary resources for innovation from both inside and outside their organization. Companies that make use of more resources will ultimately profit from the broader breadth of ideas available to them. Instead of holding onto innovations from concept stage until entry to market, organizations should learn to profit from licensing and/or selling intellectual property (IP) when it does not fit their goals or business model, and they should learn to make use of external IP when it is a good fit.

Creating an innovation reef is not simple, but it can be done and, as evidenced by ATI, can yield significant value to organizations and regions. The reef model must be used as an organizational strategy, not dictated from the top or initiated by consultants, but born out of joint vision and participation throughout the organization. Organizations and incubators must be open to the innovation process—encouraging participation and cooperation across groups or normal siloed structures. The environment must tolerate risk—which brings with it opportunities for both success and failure. In all cases, engagement, responsibility, and "pulling one's own weight" are required to form and operate successful teams.

Experienced individuals who track problems to solutions through innovative approaches must be linked with subject matter experts to develop reasonable approaches. In this context, the expert generalists play a critical role. Their broad knowledge, openness to new knowledge, and experience allow them to cut across technical and market verticals to conceptualize strategies that are both reasonable and testable. Furthermore, they tend to bring together disparate players, increasing the richness of the discourse and depth of evaluation. Managers must be trained to evaluate the proximity of an innovation strategy to market realization and assist teams to accelerate toward monetizable ends. Monetization requires connecting different parts of the business community to the innovation strategy, both to validate and cooperate, so that it can be realized.

Managers must understand that their role is not to dictate outcomes but, rather, to create structures that encourage engagement and cooperation across traditional siloes. The energy and outcomes of such a structure should encourage and recruit participation through the trading of broadly applicable currency within the organization. Value is the ultimate measure for both individual participants and the reef itself. Individuals should amplify their contributions through the synergistic power of network effects that are put in place, and the reef must create measurable outcomes providing value to the participants and the overall organization. Reef management must fight the tendencies of individuals to form or joint siloes diluting the needed resources as observed in the environment between the shore and reef. The reef must also mitigate the tendency of individuals to approach one another with a strictly competitive perspective as observed in deeper waters full of predators. Careful management of rewards and team-based value can create a vital, collaborative environment in which individuals invest in the formation of a larger organism, like a reef, and produce results beyond their separate potential.

CONCLUSIONS

We began this chapter by asking the question, "What should social structures look like to support innovation and new entrepreneurial ventures?" We have observed the importance of open structures to not only develop innovations but also bring together the right types of people to help new businesses thrive. The coral reef model conceptually organizes the social structures that are necessary to accomplish this goal. The reef provides an open structure to not only develop innovations and collaborate on their development but also to bring together the right types of people to help new businesses thrive that seek to commercialize the resulting products. In a reef, companies can more efficiently find capital to develop these novel products and launch these into the market under more efficient timelines through the effective use of resources. Outside guidance from industry experts plays a critical role in iteration of the business model and target markets, which continue to evolve in the startup process based on industry and potential customer feedback. This helps companies to not only launch their products faster but also launch better products and to do so in a more strategic manner.

REFERENCES

Aldrich, H. E., & Zimmer, C. (1986). Entrepreneurship through social networks. In D. L. Sexton & R. W. Wilson (Eds.), *The art and science of entrepreneurship* (pp. 154–167). Cambridge, MA: Ballinger.

Bollingtoft, A., & Ulhoi, J. P. (2003). The networked business incubator—Leveraging entrepreneurial agency? *Journal of Business Venturing, 20*, 265–290.

Bunderson, C. V. (2003). How to build a domain theory: On the validity centered design of construct-linked scales of learning and growth. In M. Wilson (Ed.), *Objective measurement: Theory into practice*. New York, NY: Ablex.

Chesbrough, H. (2003). *Open innovation: The new imperative for creating and profiting from technology*. Boston, MA: Harvard Business Review Press.

Chesbrough, H. (2006). *Open business models*. Boston, MA: Harvard Business School Publishing.

Etzkowitz, H., & Leydesdorff, L. (1995). The triple helix—University–industry–government relations: A laboratory for knowledge based economic development. *EASST Review, 14*, 14–19.

Freeman, C. (1995). The national innovation systems in historical perspective. *Cambridge Journal of Economics, 19*, 5–24.

Freeman, C., & Lundvall, B. Å. (Eds.). (1988). *Small countries facing the technological revolution*. London: Pinter.

Grimaldi, R., & Grandi, A. (2005). Business incubators and new venture creation: An assessment of incubating models. *Technovation, 25*, 111–121.

Hansen, M. T., Chesbrough, H. W., Nohria, N., & Sull, D. N. (2000, September–October). Networked incubators: Hothouses of the new economy. *Harvard Business Review, 78*(5), 74–84.

Henkel, T. P. (2010). Coral reefs. *Nature Education Knowledge, 3*(10), 12.

Jarrett, J. E., & Field, R. (2014, January). *The economic impact of Austin Technology Incubator and alumni companies on Travis County, 2003–2012*. Austin, TX: Bureau of Business Research, The University of Texas at Austin.

Muro, M., & Katz, B. (2010, September). The new "cluster moment": How regional innovation clusters can foster the next economy. Metropolitan Policy Program, Brookings Institute. Retrieved December 1, 2014, from http://www.wedc.wa.gov/Download%20files/2010.09-ClusterMoment-Brookings.pdf.

Porter, M. E. (1990a, March-April). The competitive advantage of nations. *Harvard Business Review*.

Porter, M. E. (1990b). *The competitive advantage of nations*. New York, NY: Free Press/Macmillan.

Ratinho, T. (2011). *Are they helping? An examination of business incubators' impact on tenant firms*. Doctoral dissertation, University of Twente Portugal. Zutphen, The Netherlands: CPI Wöhrmann. ISBN: 978-90-365-3263-1.

Saxenian, A. L. (1996). Inside-out: Regional networks and industrial adaptation in Silicon Valley and Route 128. *Cityscape: A Journal of Policy Development and Research 2*, 41–60.

Smilor, R., Gibson, D. V., & Kozmetsky, G. (1988). Creating the technopolis: High technology development in Austin, Texas. *Journal of Business Venturing, 4*, 49–67.

Smilor, R., & Gill, M. (1986). *The new business incubator: Linking talent, technology, capital, and know-how*. Lexington, MA: Heath.

Stoker, J. I., Looise, J. C., Fisscher, O. A. M., & DeJong, R. D. (2001). Leadership and innovation: Relations between leadership, individual characteristics and the functioning of R&D teams. *International Journal of Human Resource Management, 12*, 1141–1151.

Swart, P. K. (2013). Coral reefs: Canaries of the sea, rainforests of the oceans. *Nature Education Knowledge, 4*(3), 5.

Wadhwa, V. (2013, July 3). Silicon Valley can't be copied. *MIT Technology Review*. Retrieved November 25, 2014, from http://www.technologyreview.com/news/516506/silicon-valley-cant-be-copied.

Wiggins, J., & Gibson, D. V. (2003). Overview of US incubators and the case of the Austin Technology Incubator. *International Journal of Entrepreneurship and Innovation Management, 3*, 56–66.

INDEX

Note: Page numbers followed by the italicized letters *f*, *n* and *t* indicate material found in figures, notes and tables.

action planning
   in creative problem solving, 72*f*
   as execution step, 74
   synthesis in, 87
"agglomeration," 208
agreement(s)
   in brainstorming groups, 46
   in research contracts, 181
Alcan Engineered Products
   (Constellium), 171, 183–186, 196*f*
Altshuller, G., 83
ambition template, 81*f*, 84–85
analogical distance, cognitive study of
   design problem, 21
   experimental design, 20
   experimental procedure, 23
   ideation metrics, 23–24
   participants in, 21
   results, 24–26
   stimuli selection, 21–23
analogical distance, "sweet spot" study on
   conditions for, 31–32
   patent set choice, 30–31
   procedure, 32
   results, 32–38
   stimulus set choice, 31
   study participants, 31

analogical stimuli
   commonness of, 17–18
   computational design tools in, 19
   modality of, 16–17
   selection of, 21–23
analogy, use/benefits of, 14–16, 83. *See also* design-by-analogy
angel networks/business angels, 104, 206, 213, 216
Apple, 3, 206
assumptions, implicit, 154–156
Åstebro, T., 126
Austin, 5. *See also* University of Texas (UT Austin)
Austin Technology Incubator (ATI)
   coral reef structure and, 214–219
   model at, 212–214
   USAA collaboration, 125

Bain & Company, 43, 141
barriers
   clustering/incubation and, 210, 211
   to effective problem definition, 44, 72–73, 76–79
   to innovation, 203–205, 219–220
   in Nestlé example, 192
   to OI implementation/adoption, 143, 156, 159–160

barriers (*Cont.*)
   in organizational structures, 101
   Quicklook process and, 126
   removal of, 152
   to success, 18
   in why-why-why analysis, 80*f*
Baruah, J., 50, 56
Basadur, Min, 71, 77–79
Bayesian model
   in human cognition, 26
   for structural form discovery, 27, 29
Bayh-Dole Act, 104, 206
Bell Labs, 6, 104, 206
Bezos, Jeff, 147
Boone, Todd, 145
Boston, as high-tech hub, 2–3, 210
brainstorming groups. *See also* creative group processes
   brainwriting in, 52–55, 58, 62
   breaks in, 49–50
   diversity of experience in, 56–57
   effectiveness of, 50–51
   facilitation/training in, 55–56
   group/individual alternation, 53–55
   idea selection, 59–60
   key processes/task factors in, 45*b*
   leadership role/styles, 60
   limitations of, 8–9
   in organizational settings, 61
   personality characteristics in, 57–58
   recommendations for, 61–62
   research directions, 62–63
brainwriting, 52–55, 58, 62. *See also* brainstorming groups
brands/brand management, 101
Bureau for Business Research, 125
business angels/angel networks, 104, 206, 213, 216

California Institute of Technology (Caltech), 177, 199*n*3
Cambridge University, 177, 199*n*3
candle problem (Duncker), 17
capability building
   individual development, 158–159

   internal connection facilitation, 159–161
   investment in, 157–158
causal chains, 96, 101–102, 111
Central Texas Angel Network, 216. *See also* business angels/angel networks
challenge statement, 74–75, 78, 83
Chesbrough, H. W., 1, 210
"chunk," ability to, 18
Cisco, 3, 171, 182, 188–190, 196*f*
cluster/clustering concept, 209–211
co-creation, 114
cognitive study, on design-by-analogy
   design problem, 21
   experimental design, 20
   experimental procedure, 23
   ideation metrics in, 23–24
   participants in, 21
   patent selection, 21–23
   results of, 24–26
cognitive styles, 78
collaborative process. *See* brainstorming groups; creative group processes
commercialization, of products
   acceleration of, 109
   ATI and, 215, 219
   consumer segments and, 113
   ecosystem interaction, 170, 170*f*
   forecasts and, 126
   incubator function and, 212
   innovation cell competencies and, 172*f*
   innovation support for, 1, 105
   OI application to, 121
   Pole Cam project, 137–138
   technology transfer and, 184
Commission for Technology and Innovation (CTI), 177, 196*f*
common properties
   of analogical stimuli, 17–18
   cognitive study of, 20–26
communication, of vision, 146–148
competitive advantage, 141, 174, 198, 209
The Competitive Advantage of Nations (Porter), 209

Index

complexity, 26, 43, 75, 195, 207
components
 of innovation, 205
 in 9 Windows analysis, 83*f*
 for Pole Cam, 132–133
 rapid improvement in, 4–5
computation support, for
 design-by-analogy
 LSA in, 27–28
 methodology, 27
 patent organization, 26–27
 structural form discovery in, 28–30
"conceptualizers," 78
Connect & Develop, 7, 78, 174. *See also*
 Procter & Gamble (P&G)
ConocoPhillips, 151
Constellium (Alcan Engineered
 Products), 171, 183–186, 196*f*
coral reef model
 ATI comparison with, 214–219
 as barrier solution, 219–220
 concept of, 204
 creation of, 220–221
Cornwell, B., 126
creative group processes. *See also*
 brainstorming groups
 "natural meetings," 44, 46
 number of ideas, 47–50
 theoretical models, 46–47
creative problem solving
 action planning and, 87
 crowdsourcing and, 74, 84–85
 customer/supplier engagement, 85–87
 fact-finding in, 86
 idea evaluation/selection in, 87
 idea finding in, 86
 open innovation vs., 75–76
 opportunity finding in, 85
 problem definition in, 72–73,
  76–81, 86
 process for, 72*f*
 recommendations for, 87–88
 toolbox for, 81–84
 why-why-why analysis, 80*f*
Credit Suisse, 193–195

cross-functional/-disciplinary teams. *See
 also* multidisciplinary networks
 complexity and, 207
 at Constellium, 196*f*
 internal capabilities in, 158
 leaders' role in, 160
 opportunity thinking approach in, 93
 reporting functions, 173*f*
crowdsourcing, 74, 84–85
customer/supplier engagement,
 74–75, 85–87

Dahl, D. W., 15
Dahmén, Eric, 208
decision making, 44, 105, 126, 220
De Dreu, C. K. W., 58
definition, of problem, 72–73,
 75, 76–81
Dell, 6
Departamento de Ciência e Tecnologia
 Aeroespacial, 177
design-by-analogy
 analogical distance in, 15–16
 cognitive study of, 20–26
 computational support for, 26–30
 described, 13
 open goals in, 14
design fixation, 14
design thinking/design thinking firms
 in creative problem solving, 75–76
 in innovation process, 72
 methods/approach of, 8–9
design tools, 19
development blocks, 208
Digital Equipment Corporation, 3
digital image capture, 128. *See also* Pole
 Cam project
disruptive innovation
 barriers to, 204
 Cisco and, 189
 Constellium and, 183–186
 EPFL's Innovation Square and,
  176–181
 failure of, 3
 ideation and, 7–8

disruptive innovation (*Cont.*)
  in innovation cells, 169, 171, 173–176, 197–198
  market opportunity and, 99
  Nestlé and, 192
  Peugeot Citroën and, 186
  support of, 4
diversity of experience, 56–57. *See also* brainstorming groups
Drucker, Peter, 154
drug companies, 6
DSM (science-based company), 73, 147
Duncker, K., 17
DuPont, 104, 156, 206

East Coast, innovation strategies on, 2–4, 210
Eastman Kodak (Kodak), 95, 107–110, 116
École Polytechnique Fédérale de Lausanne (EPFL). *See* EPFL
ecosystem insight, 111–112. *See also* opportunity thinking
ecosystem(s)
  coral reef model, 177, 215–216
  differences in, 2–6
  at EPFL, 170–171
  external, 103–106
  history of, 207–211
  internal, 106–107
  open innovation as, 211–212
  in start-ups, 169, 172*f*, 173*f*, 197
Einstein, Albert, 76
electronic brainstorming, 51–52, 60, 62. *See also* brainstorming groups
Elephant in the Dark (Rumi), 92*f*
Embraer (Brazil), 177
employee attitude/motivation, 144–145
employee capability building
  individual development, 158–159
  internal connection facilitation, 159–161
  investment in, 157–158
entrepreneurial alertness, to opportunity, 94–98

environment, as opportunity source, 102. *See also* opportunity/opportunities
EPFL
  Cisco and, 188–190
  Constellium and, 183–186
  Credit Suisse and, 193–195
  innovation ecosystem of, 170–171
  Innovation Square and, 176–181
  Nestlé Institute of Health Sciences (NIHS) and, 190–193
  Peugeot Citroën Group and, 186–188
ETH (Switzerland), 177, 199*n*3
examples, introduction of, 14. *See also* design-by-analogy
experience filter, 97, 112
expression, as opportunity source, 101. *See also* opportunity/opportunities
external ecosystem, 103–106. *See also* ecosystem(s)
extraversion, in brainstorming groups, 58. *See also* brainstorming groups
Exxon Valdez oil spill, 84

Facebook, 156
facilitation/training, in brainstorming, 55–56. *See also* brainstorming groups
failure/failed ventures
  as acceptable, 157
  career damage from, 210
  compensation impact, 204
  Drucker on, 154
  in entrepreneurial process, 4
  "heroic failure" award, 157
  as inhibiting process, 45*b*
  in R&D thinking, 155–156
  risk spreading/tolerance, 5, 221
feedback, 48. *See also* brainstorming groups
Fiat, 159, 173
financial incentives, 156–157
financial resources, for OI, 152
Fisher, W., 78, 81
5 Why Analysis, 80–81

Flash Dry technology (The North Face), 113–114
formal/informal interactions, 180–182
40 Inventive Principles (TRIZ), 19, 77, 83
Freeman, Christopher, 208
functional analysis, 82
functional diversity, 57, 63. *See also* brainstorming groups
functional fixedness, 18, 97
functional modeling, 19
funders, 211, 220. *See also* business angels/angel networks

Gelade, G., 78
Gibson, D. V., 208
Gielnik, M. M., 98
Gladwell, Malcolm, 9
Global Commercialization Group, 125
Goldenberg, O., 53
Grey Advertising, 157
group idea generation, 8. *See also* brainstorming groups
GYM (P&G innovation center)
  culture example, 71–72
  effectiveness of, 77–79
  idea generation at, 8
  problem solving workshops, 74

Hansen, M. T, 210
Hawker, Lizzie, 113
"heroic failure" award, 157
heuristics, 19, 96–97, 112, 134
Hewlett-Packard, 3, 142
hierarchy structure, 34–36
history, of innovation theory, 207–211
homeowner insurance claims. *See* Pole Cam project
hubs
  cluster/clustering concept, 209–211
  regional differences, 2–6
human resources, for OI, 152

IBM, 3, 104, 142, 149
IC^2 Institute, 36, 121–122, 125–126, 134
ICorp program (National Science Foundation), 125
idea scouts/brokers, 159, 160–161
ideas/ideation
  ability to "chunk," 18
  building on, 48
  innovation and, 7–9
  interventions for, 47–48
  metrics for, 23–24
  vs. opportunities, 95
  quality of, 47, 56, 59, 62
  selection of, 59–60
idea-to-product development, 204
IDEO (design thinking firm), 8–9
Imperial College, 177
implementation, of OI
  capability building, 157
  commitment to, 150–151
  development of individuals, 158–159
  implicit assumptions, 154–156
  internal connection facilitation, 159–161
  reward alignment, 156–157
  shared ownership, 148–150
  support for, 151–154
  vision crafting/communication, 146–148
"implementers," 78
incremental innovation, 3, 174–176, 183, 189, 198
incubators. *See also* Austin Technology Incubator (ATI)
  at MIT, 200n4
  networked, 6, 210–212
  region-specific, 209
  social structures and, 203
India, PepsiCo expansion into, 111–112
individuals
  in brainstorming groups, 57–58
  capability building of, 158–159
Industrial Design Society of America, 109–110
informal/formal interactions, 180–182
InnoCentive, 76

innovation. *See also* implementation, of OI; open innovation (OI)
  academic discussions of, 10
  barriers to, 204–205
  as business buzzword, 1
  definition of, 204
  democratization of, 104
  evaluation process, 9–10
  ideation and, 7–9
  in organizational settings, 61
  as priority, 43
  toolbox for, 81–84
innovation cells. *See also* disruptive innovation; Innovation Square
  competencies of, 172*f*
  ecosystem interaction, 170*f*
  engineering technology universities, 176–177
  formal/informal interactions, 180–182
  introduction to/description of, 169–172, 197–198
  at large firms, 182–183
  R&D/BU use of, 172–176
  roadmap for, 173*f*
Innovation Communities for Enterprise (ICE), 123–124
Innovation Square
  Cisco and, 188–190
  collaboration examples at, 196*f*
  Constellium and, 183–186
  Credit Suisse and, 193–195
  ecosystem of, 170–171
  at EPFL, 176–181
  formal/informal interactions, 180–182
  Nestlé Institute of Health Sciences (NIHS) and, 190–193
  Peugeot Citroën Group and, 186–188
innovation strategies. *See also* design-by-analogy
  ecosystem differences, 2–6
  minimal evaluation approach, 48
  proximity to market, 221
innovation studios. *See* GYM (P&G innovation center); IDEO (design thinking firm)

innovation techniques, transfer of, 1. *See also* technology transfer
innovative ideas, reactions to, 43–44
Innovator Certification (USAA), 122, 126, 133–137
Instituto Tecnológico de Aeronáutica, 177
insurance claims. *See* Pole Cam project
intellectual property, 100, 171, 178, 181, 187, 189, 199n2, 220
interaction modalities, 51. *See also* brainstorming groups
interactions, formal/informal, 180–182
internal connection, facilitation of, 159–161
internal ecosystem, 106–107. *See also* ecosystem(s)
Internet, 85, 104–105, 190, 206
introversion, in brainstorming groups, 58. *See also* brainstorming groups
inventors, innovation by, 2
iPad, 124, 133, 206
iPhone, 124, 206
iterative practice/process, 72, 74, 75

Jansson, D. G., 14

Kahneman, Danny, 9
Kemp, C., 26, 27, 30
Kennametal, 73–74
Kodak (Eastman Kodak), 95, 107–110, 116
Koehler, D. J., 126
Kozmetsky, George, 125, 208
KTH Royal Institute of Technology, 177
Kumar, K., 56

Lackner, Tomas, 150, 157
Lafley, A. G., 71–72
landscaping process, in opportunity thinking, 111–112
large companies. *See* Austin Technology Incubator (ATI); brainstorming groups; EPFL; R&D
Larson, J. R., 53

Index

Latent Semantic Analysis (LSA), 27–28
Lawler, E. E., 149
leadership roles/leader behaviors
  in brainstorming, 60
  capability building, 157
  commitment, 150–151
  development of individuals, 158–159
  implicit assumptions and, 154–156
  internal connection facilitation, 159–161
  reward alignment, 156–157
  shared ownership, 148–150
  support, 151–154
  vision crafting/communication, 146–148
Leydesdorff, Loet, 209
Linsey, J. S., 55
List, George Friedrich, 207–208
Lundvall, Bengt Ake, 208

Maier, N. R. F., 17–18
market, as opportunity source, 99. See also opportunity/opportunities
market results, in opportunity thinking, 114
Markman, Art, 78
Marshall, Alfred, 208
MCC Consortium, 5
McDonald's (hot coffee lawsuit), 82
McKinsey Global Survey, 43
means-end framework, 96, 99, 101
Merck, 104, 182, 206
Michaelson, L. K., 56
minimal evaluation approach, 48. See also brainstorming groups
MIT (Massachusetts Institute of Technology), 3, 177, 199n3, 199n4
mobile apps/devices, 124
modality
  of analogical stimuli, 16–17
  cognitive study of, 20–26
  interaction modalities, 51
monetary/non-monetary rewards, 156–157
Moore's law, 4

Moreau, P., 15
motivational basis, for creativity, 47. See also brainstorming groups
multidisciplinary networks.
  See also cross-functional/-disciplinary teams
  Constellium example, 184, 196f
  in innovation cells/ecosystems, 198, 207
  at MIT, 199n4
  in organizational structures, 174
  physical collocation of, 176
  Rolex Learning Center example, 178
Myhren, Tor, 157

NASA Mid-Continent Technology Transfer Center, 126
National Science Foundation (NSF), 125
"national system of innovation," concept, 208
"natural meetings," 44, 46. See also brainstorming groups
Nelson, B. A., 15
Nestlé Institute of Health Sciences (NIHS), 190–193
networked incubators, 6, 210–212. See also Austin Technology Incubator (ATI)
Networks of Excellence, 151
NIH (not-invented-here) syndrome, 159–160, 190
9 Windows analysis, 82, 83f
NineSigma, 76
Nohria, N., 210
The North Face, 113–114, 116
not-invented-here (NIH) syndrome, 159–160, 190
novel ideas/idea novelty
  analogical stimuli and, 13–18, 33f, 36
  business incubators and, 209–210
  coral reef concept and, 222
  engineering universities and, 177, 197
  entrepreneurial experience and, 96–97
  far-field examples and, 24–25, 32
  group ideation and, 48–49, 54

novel ideas/idea novelty (*Cont.*)
  idea selection and, 59–60
  organization approach to, 174
  personality characteristics and, 58
  risk and, 204
  in startup operations, 220

open goals, 14. *See also* design-by-analogy
open innovation (OI). *See also* implementation, of OI; innovation; leadership roles/leader behaviors
  competitive advantage, 141
  vs. creative problem solving, 75–76
  crowdsourcing and, 74, 84–85
  customer/supplier engagement, 74–75, 85–87
  as ecosystem, 211–212
  embedding of, 162
  external ecosystem, 103–106
  forces driving, 2–6
  internal ecosystem, 106–107
  micro-level perspective on, 143–145
  opportunity thinking synergies, 116
  paradigm adoption, 142
  purpose of, 7
  as term, 1
  toolbox for, 81–84
  transition to, 205–207
  trends in, 73–74
open organizational structures, 173–176, 222
opportunity finding, in creative problem solving, 85
opportunity/opportunities
  entrepreneurial alertness to, 94–98
  vs. ideas, 95
  selection/stretching of, 112–113
  sources of, 98–103
  term usage, 94
opportunity thinking
  co-creation in, 114
  as conceptual tool, 115–116
  Eastman Chemical Company example, 107–110

  ecosystem insight in, 111–112
  "elephant" metaphor for, 91–94
  market results and, 114
  selection/stretching of, 112–113
organization, as opportunity source, 101. *See also* opportunity/opportunities
Osborn, A. S., 8, 48
Outstanding Corporate Innovator (award), 73
ownership, promotion of shared, 148–150
Oxford University, 177, 199*n*3

parametric analysis, 82
patents. *See also* US Patent database/classification system
  organization of, 26–27
  selection of, 21–23
  use of, 19
Paulus, P. B., 50, 51, 53, 54, 56
peer facilitation, 78–79
PepsiCo, 111–112
personality characteristics, in creative group processes, 57–58. *See also* brainstorming groups
Peugeot Citroën Group, 186–188
pharmaceutical companies, 6
Philips, 152
pictorial-based representations, 16–17
Pink, Dan, 9
Pole Cam project
  commercialization of, 137–138
  field trials of, 129–131
  idea vetting, 128–129
  manufacturing/training, 132–133
  as open innovation success, 136
  pictures of, 132*f*
  problem definition, 127
  Quicklook process and, 133–135
  research partnership for, 136–137
  stakeholder response, 131–132
  UAV use, 127–128
Porter, Michael, 209
Portugal, open innovation in, 94
problem definition

barriers to, 44
in creative problem solving, 72–73, 75–81, 86
in Pole Cam project, 127
workshops for, 78–79
problem solving. *See* creative problem solving
Procter & Gamble (P&G)
Connect & Develop, 7, 78, 174
GYM innovation center, 8, 71–72, 74
innovation tools at, 81
OI paradigm adoption, 142
OI team effectiveness, 77–79, 152
training/development at, 159
vision communication, 147
Product Development and Management Association, 73
production blocking, 45b, 46, 50–52. *See also* brainstorming groups
prosocial motivation, in brainstorming groups, 58. *See also* brainstorming groups

quadruple-helix model, 209
quality
of ideas, 47, 56, 59, 62
of solution concepts, 16, 24–25, 32–33
quantity goals, in brainstorming, 48. *See also* brainstorming groups
Quicklook method, 126–128, 133–137

R&D
business unit interaction, 170f
failure/failed ventures and, 155–156
vs. innovation cells, 172–176
of large companies, 6–7
Raytheon, 3
reaction, to innovative ideas, 43–44
research universities, 5, 21, 104, 206. *See also specific universities*
reward systems/alignment, 101, 104, 154, 156–157, 206, 221
risk financing, 104, 206
risk spreading/tolerance, 5, 221
Roche Diagnostics, 156

Rolex Learning Center, 178, 179f, 189
roof inspection process. *See* Pole Cam project
Rosen, D., 15
Rumi, 91–93

"safe zone," 218, 220
Saxenian, A. L., 2–3, 210
Segway, 10
shared ownership, 148–150
Siemens, 150, 152
Silicon Valley
as cluster example, 209–210
ecosystem of, 4–6, 211
lessons from, 7, 11
rise of, 2
Simplex model (Basadur), 71, 72f, 79
Six Sources of Opportunity, 98–103, 107–108, 110, 112, 115–116
Smilor, R., 208
Smith, S. M., 14, 49
social capital, 161, 211, 216, 219
social events/gatherings, 181, 182f, 198
social structures, for innovation support, 203–205. *See also* coral reef model
solution concepts, metrics for, 23–24
solution transfer, 24, 25. *See also* technology transfer
space management, 76
Spradlin, D., 83
Stanford, 2, 177, 199n3
startups
ATI and, 211–212, 215–216
emergence of, 3
entrepreneurial alertness and, 101
gatherings for, 181
innovation barriers for, 204
in innovation cells/ecosystems, 169, 170f, 172f, 173f, 197
Innovation Square and, 177–178, 182, 186, 195, 196f
model/process for, 220, 222
Nestlé's use of, 192
organization approaches to, 174
in Silicon Valley, 4–6, 210–211

Stone, R., 19
strategic design. *See* design-by-analogy
structural form discovery/types, 27–30
success, meaning of
   implicit assumptions, 154–156
   reward alignment in, 156–157
Sull, D. N., 210
supplier/customer engagement, 74–75, 85–87
Swiss Federal Institute of Technology. *See* EPFL
Switzerland, 177. *See also* EPFL

task structure, 50. *See also* brainstorming groups
Technical University of Munich, 177
technology, as opportunity source, 100–101. *See also* opportunity/opportunities
technology hubs
   cluster/clustering concept, 209–211
   regional differences, 2–6
technology transfer
   Credit Suisse example, 194
   engineering universities and, 197
   at IC^2 Institute, 125–126
   from innovation cells to BUs, 171–173, 184
   Peugeot Citroën example, 187
   in research contracts, 181
   in Switzerland, 177
technopolis model, 5, 208–209, 212
Tenenbaum, J. B., 26, 27, 30
Texas Instruments, 157
text-based representations, 16–17
Blind Men and the Elephant (Saxe), 91, 93
3M, 149, 160, 185
time, as OI resource, 152
Timmons, A., 94, 99
tools, for innovators, 81–84
training/facilitation, in brainstorming, 55–56. *See also* brainstorming groups
transfer. *See* technology transfer

transformational leadership, 60
triple helix model, 5, 209
TRIZ (theory of inventive problem solving), 19, 77, 83
trust networks, 218
Tseng, I., 14
two-string problem (Maier), 17–18

Unilever, 152, 159
United States
   business incubators and, 211
   facial recognition technology use, 111, 188
   Nobel Prize winners, 199n4
   Pole Cam rollout throughout, 132
   technology company development, 2
   technology/engineering universities, 177
University of Texas (UT Austin). *See also* Pole Cam project
   coral reef/ecosystem concept, 177, 211–212, 215–216
   in high-tech hub success, 5
   USAA collaboration, 121–122, 125–127
unmanned aerial vehicles (UAVs), 127–128. *See also* Pole Cam project
USAA (United Services Automobile Association). *See also* Pole Cam project
   innovation at, 123–125
   mission/structure of, 122–123
   UT Austin collaboration, 125–127
US Patent database/classification system, 21, 26, 29, 30, 79

value curves, 83
value proposition, 85, 210
venture capital/capitalists
   ATI example, 213
   in coral reef concept, 177
   in early stage finance, 104, 206
   as ecosystem actors, 169, 170*f*
   innovation cell competencies and, 172*f*

in innovation cell interaction, 173f, 197
vision, crafting/communication of, 146–148

Ward, T. B., 14
Watson, W. E., 56
West Coast, innovation strategies on, 3–4, 210
"why-what's stopping" tool, 82
why-why-why analysis, 80f, 82
Wiley, J., 53
Wilson, J. O., 15
Wood, K., 19
workshops
  fact-finding in, 84, 86
  functional analysis in, 82
  for innovation, 71
  participants in, 88
  for problem definition, 78–79
  for problem solving, 74, 76
Worley, C. G., 149

Yang, H., 53, 54, 56
Yen, J., 15

www.ingramcontent.com/pod-product-compliance
Ingram Content Group UK Ltd.
Pitfield, Milton Keynes, MK11 3LW, UK
UKHW041451180426
11946UKWH00014B/158/J